BIG IDEAS *from* Dr. Small

CREATING A COMFORT ZONE FOR TEACHING MATHEMATICS

GRADES K–3

REVIEWERS

Cathy Canavan-McGrath, *South Slave Divisional Education Council*, NWT
Anna Jupp, *Toronto District School Board*, ON
Manuel Silva, *Winnipeg School Division*, MAN
Cynthia Tulissi, *Calgary Catholic School District*, AB

NELSON PROFESSIONAL LEARNING

NELSON EDUCATION

Big Ideas from Dr. Small
Creating a Comfort Zone for
Teaching Mathematics Grades Kindergarten–3

Author
Marian Small

General Manager, Mathematics, Science, and Technology
Lenore Brooks

General Manager, Marketing, PRIME
Audrey Wearn

Publisher, Mathematics
Colin Garnham

Associate Publisher, Mathematics
Sandra McTavish

Managing Editor, Development
Erynn Marcus

Product Manager
Linda Krepinsky

Program Manager
Caron MacMenamin

Project Manager
Alisa Yampolsky

Developmental Editor
Jackie Williams

Assistant Editor
Sarah Brown

Editorial Assistant
Kathryn Chris

Senior Content Production Manager
Sujata Singh

Content Production Editor
Shana Hayes

Copy Editor
Carolyn Wagner

Proofreader
Cheryl Tiongson

Indexer
Noeline Bridge

Production Manager
Helen Jager-Locsin

Production Coordinator
Susan Ure

Design Director
Ken Phipps

Interior Design
Peter Papayanakis

Cover Design
Ken Phipps and Rocket Design

Cover Image
Veer.com

Asset Coordinators
Renée Forde
Suzanne Peden

Compositor
Nesbitt Graphics, Inc.

Printer
Courier Co.

Reviewers
The authors and publisher gratefully acknowledge the contributions of the following educators:
Cathy Canavan-McGrath, South Slave Divisional Education Council, NWT
Anna Jupp, Toronto District School Board, ON
Manuel Silva, Winnipeg School Division, MAN
Cynthia Tulissi, Calgary Catholic School District, AB

ISBN-13: 978-0-17-610555-6
ISBN-10: 0-17-610555-7

Printed and bound in USA
1 2 3 4 12 11 10 09

For more information contact Nelson Education Ltd., 1120 Birchmount Road, Toronto, Ontario, M1K 5G4. Or you can visit our Internet site at http://www.nelson.com

Table of Contents

Acknowledgments

We are fortunate in Canada to have ministries of education and teachers of mathematics who are open to progressive thinking about mathematics education. Because of this, Canadian teachers are willing to tackle learning more about mathematics in order to help their students.

I would like to thank those many committed educators for their interest in what I have shared with them over the years about how to make mathematics more meaningful to them and their students.

I would also like to thank Nelson Education Ltd. for being so supportive of my work and showing so much faith in me over the course of our many projects together. This book is due, in no small part, to the encouragement and efforts of a number of people at Nelson, particularly Audrey Wearn, director of professional learning, and my editor, Jackie Williams. Audrey is a consummate professional, savvy, easy to work with, and full of integrity. Jackie's keen eye and thoughtful approach ensures that what we produce is clear, coherent, and helpful to teachers. I would also like to acknowledge Caron MacMenamin, who worked with the editor, to create a lovely, accessible design and high quality DVD.

I would also like to thank the reviewers for their advice, helpful comments, and suggestions.

About the Author

Marian Small is the former Dean of Education at the University of New Brunswick. She has been a professor of mathematics education for over 30 years and is a regular speaker on K to 12 mathematics throughout Canada and the United States.

Over the last 20 years, she has been an author on seven student text series at both elementary and secondary levels in Canada, the United States, Australia, and the country of Bhutan, and a senior author on five of those series. She has served on the author team for the National Council of Teachers of Mathematics Navigation series, PreK to 2; for four years, she was the NCTM representative on the Mathcounts question writing committee for middle school mathematics competitions throughout the United States. She is also a member of the editorial panel for the NCTM 2011 yearbook on motivation and disposition.

She has recently completed a text for university pre-service teachers, *Making Math Meaningful for Canadian Students: K–8*; a book on differentiated instruction in mathematics K to 8, *Good Questions: Great Ways to Differentiate Mathematics Instruction*, published by Teachers College Press; and the companion book to this one, *Big Ideas from Dr. Small, Grades 4 to 8*.

She has led the research resulting in the creation of maps describing student mathematical development in each of the five mathematical strands for the K to 6 level and created the associated professional development program, PRIME. She has also developed materials and provided consultation for working with struggling learners and asking good questions.

Introduction

Issues in Teaching Mathematics

Mathematics at the K to 3 level is not just about counting, adding, and subtracting. The curriculum is much richer than that. Because the mathematics curriculum has become increasingly sophisticated over the last number of years, many teachers do not have the benefit of a grounding in some of the content they must teach. Even where the content is familiar, the approaches to teaching mathematics that the curriculums across the country require are a far cry from what many teachers have experienced themselves. In other words, teaching math, even to young students, can be complex and challenging.

The mathematics and teaching approaches have become more sophisticated over the years.

The Power in Being Comfortable with the Math We Teach

The question is, what can we do to make the challenge more manageable and productive? It might help to look at a metaphor. All of us know from the many facets of our lives how much more successful we are in accomplishing tasks when we are confident and comfortable. We also know that that air of confidence is usually sensed by those around us and often results in a level of comfort for them as well.

For example, when you drive visitors around your neighbourhood, you have an intuitive sense of what to do and your passengers are confident that you have the background knowledge that will guide you to your destination. However, when you take them to locations that are unfamiliar to you, they sense that you are tentative and reliant on "rules" (in this case maps) to which you repeatedly refer. So, unlike when you are in your comfort zone where your passengers are at ease, your passengers again take their cues from you, except this time they are uneasy about how, and if, they will reach their destination.

Similarly, our students are our "passengers." Our job is to build their confidence and reassure them that they really can understand the new ideas with which they are dealing. This is accomplished by helping them create an internal map of how these new ideas connect to what they already know. This means we, as teachers, need that internal map, too, in order for our students to recognize our comfort and confidence.

Our job as teachers is to imbue our students with the confidence they need to understand the new ideas with which they are dealing.

What Teaching to Big Ideas Means

One of the gaps in many teachers' mathematical backgrounds is an internal map of the subject. They lack a fundamental understanding of how various mathematical topics interconnect, which topics are more important in the long term than others, and which aspects of those topics are most important. As they teach, many teachers often feel like they are going through a

Using a checklist of discrete outcomes/expectations to teach is in stark contrast to what we know from research about how students learn most effectively by making connections.

checklist, checking off whether students have learned each new discrete concept or skill listed in the curriculum. This is in stark contrast to what we know from research about how much more effective it is for students to learn when connections are explicitly made between new knowledge and ideas that students already know (Borko and Putnam, 1995; Schifter, Bastable, and Russell, 1997; Kennedy, 1997). Big ideas help teachers and students make these connections.

For example, it is important, as each operation is introduced and developed, that teachers help students make connections among the operations. So, instead of introducing multiplication as a completely new operation, it can be introduced as a "short cut" for addition that works in situations where equal groups are involved. This is the focus of Big Idea 1 for Whole Number Operations on page 35. (There are relationships between the four operations.)

Another challenge for teachers is interpreting their curriculum. Big ideas can help here as well. Some curriculum outcomes/expectations are broader and more encompassing than others, but it is not obvious when you look at the curriculum. Using big ideas can help teachers assess the attention or emphasis required to achieve various outcomes/expectations. Big ideas also helps the teacher understand what some of the broader outcomes/expectations mean. For example, suppose the curricular goal is "estimate and measure length, height, and distance using non-standard units." It may not be immediately apparent what we want students to learn while working with non-standard units. The Big Ideas for Measurement on page 85 can help with this. When students work with non-standard units, students can explore things such as

- the same object can be described using different measurements (Big Idea 1)
- any measurement can be determined in a variety of ways (Big Idea 2)
- the choice of unit affects the numerical value of the measurement—the bigger the unit, the fewer units needed (Big Idea 6)

If a teacher understands that big ideas like these are behind this expectation/outcome, he or she can teach the concepts and skills involved much more meaningfully instead of a more superficial skills-based approach.

Another way that big ideas can help is to ensure teachers and students understand the overriding purpose behind a lesson or task. Teachers often use tasks that appear to be successful because they are very engaging to students. However, the reason why the students are doing the task is not clear and the teacher does not know what to focus on in his or her questioning throughout the task and at the conclusion of the task. Teaching through big ideas is about teachers looking critically at a task or lesson, asking themselves why they are teaching that lesson or task, and then making sure that purpose becomes clear to students so that the task or lesson can be more effective. For example, many Kindergarten to Grade 2 teachers have students engage in shape hunts without realizing that the focus is not so much on identifying and naming shapes but on recognizing that some geometric attributes are similar for 2-D shapes and 3-D shapes and that some attributes are quantitative, while others are qualitative (Big Ideas 1 and 2 for Shapes and Their Properties on page 63).

There are always many ways to sort or organize data (Big Idea 2 for Organizing and Collecting Data on page 115). The use of sorting materials such as attribute blocks can help bring out this big idea. For example, this set of attribute blocks could be sorted by colour, size, thickness, or shape.

Despite increasing abstraction as students go up the grades, and shifts in focus from grade to grade, big ideas remain fundamentally simple and are often the same from grade to grade. For example, the second Big Idea for Patterns on page 1, that the mathematical structure of a pattern can be represented in a variety of ways, applies whether students are creating simple shape patterns or more complex numerical patterns. Another example is the notion that it is important not only for students to read information from graphs but to make inferences and draw conclusions (Big Idea 3 for Displaying and Analyzing Data on page 122). This big idea applies to concrete graphs in Kindergarten and Grade 1, to bar graphs in Grade 3, and to histograms in Grade 8.

Big ideas remain fundamentally simple and are often the same from grade to grade through the elementary school system.

What is most important, though, is that the big ideas not be implicit, but explicit. Whether it is through the task itself, through teacher questioning and prompting, or through teacher talk, it is important that big ideas be said out loud. The more students hear an idea, the more likely it is that they will internalize it and be able to use it to support further learning. For example, within the Number strand, Big Idea 3 on page 15 states that you can represent a number in a variety of ways and that each representation can focus on a different aspect of that number. While students are learning to represent the number 6, a question such as "How could you represent 6 to show that it is greater than 5?" would bring that big idea to the surface.

It is important to teach not only through big ideas but to speak about the big idea explicitly.

When students work with 100 charts, the following two big ideas can be made explicit through the types of tasks assigned and through effective questioning:

- Some ways of displaying data highlight patterns (Big Idea 3 for Patterns).
- The place value system we use is built on patterns to make our work with numbers more efficient (Big Idea 2 for Greater Whole Numbers).

What is most important for us to assess is student knowledge of the enduring understandings and the big ideas (Wiggins and McTighe, 1999).

A teacher should also plan both instruction and assessment by considering big ideas. For example, in planning a unit of instruction for geometry for Grade 2, a teacher might consider which of the big ideas are relevant to mathematical content in that unit. The teacher would then plan instruction to ensure that each day he/she asks questions that elicit an explicit discussion of those big ideas. In planning assessment for the unit, the teacher would develop appropriate assessment tasks around the relevant big ideas. Too often our assessment focuses on the details and often the less important aspects of the mathematics when it is the big picture that really matters.

The Purpose of This Resource

The purpose of this resource is to enhance the delivery and assessment of units of instruction. It is not meant as a full course of study, but, rather, as a tool to help you interpret your curriculum and then shape instruction and assessment in order to ensure that the most important concepts in your curriculum are made explicit and clear to your students.

Just as one would generally not teach a curriculum in its order of presentation, one does not teach big ideas in any specific order. It is simply that, in the course of teaching a curriculum in a meaningful and practical way, whether it is how a text resource does it, how a board or district has agreed to do it, or how a single teacher chooses to do it, the teacher keeps returning to the big ideas by choosing and shaping tasks and asking questions that keep bringing the big ideas to the forefront.

How This Resource Is Organized

This book is organized to focus on the big ideas in Kindergarten to Grade 3 mathematics. Each chapter focuses on one of the five mathematical strands and each chapter is then divided into familiar topics. For each topic in each strand, a set of big ideas that underlie the curriculum is articulated. For example,

Chapter 2 Number and Operations
Topics: Early Number
Greater Whole Numbers
Whole Number Operations
Fractions and Decimals

The main body of text helps the teacher better understand the mathematics that they need to know in order for their students to be successful in that strand.

Teaching Ideas are tasks and accompanying questions related to the mathematics being discussed that the teacher can use to bring the big ideas to the surface.

A special feature of the book is the Teaching Ideas in the margins. These include tasks and strategies related to the mathematics being discussed in the main body beside them, along with questions that the teacher can ask to bring the big ideas to the surface. These Teaching Ideas are intended as model tasks with model questions to elicit big ideas and many are adaptable to similar situations at another grade level that might be of more interest to a particular reader.

The example at the top of the next page from page 2 will help you understand how the Teaching Ideas are structured. Notice that specific grade levels have not been assigned to the Teaching Ideas. This is because specific content is taught at different grade levels across the country.

Chapter number

Task number

Teaching Idea **1.2**

Task

Place coloured beads in a row to make a pattern, repeating the core (e.g., blue-red) four or five times. Cover the beads with a paper tent and reveal some of the pattern as shown below. Ask students to predict what they will see next as you slide the tent over.

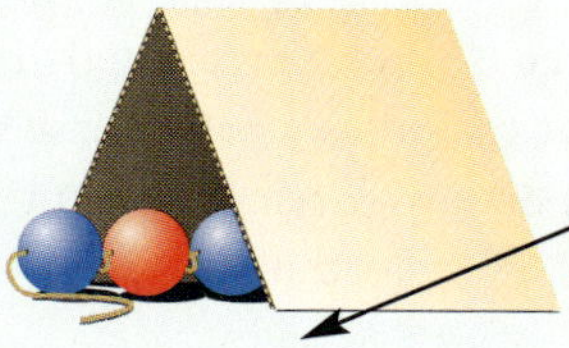

Big idea code:
BIP 1 means
Big **I**dea **1** for **P**atterns

Suggested questions the teacher can ask to bring out the big idea

To focus on BIP 1, ask: *Why did I need to tell you it was a pattern for you to predict the next bead?* [If it were not a pattern, the next bead could be any colour at all.] *Can you be sure what comes next? Why not?* [No; I don't know whether the pattern repeats blue-red or blue-red-blue.]

Sample responses

Beside Teaching Idea 1.2 in the main body of the text, the discussion focuses on repeating patterns.

Because the relevant content is discussed right beside the Teaching Idea, the reader should have no difficulty recognizing how the Teaching Idea relates to the content they teach and should be able to judge its appropriateness for their grade level.

Other Features of This Resource

In many, but not all, of the chapters, there are lists of principles or important points that will help the teacher better organize instruction. These principles and important points are generally not as broad as the big ideas, but they are important concepts to explicitly address in instruction, points that may not be obvious to all teachers.

For example, the list of Numeration Principles on page 28 refers to several specific ideas about place value. One example is that the place value system requires 0 as a place holder. This idea is not as broad as the big idea, that the place value system is built on patterns (Big Idea 2 on page 27), but it will help the teacher teach more effectively to address that big idea. Similarly, the list of Important Points About Pictographs (page 126) helps the teacher know where to focus instruction in order to address Big Ideas 1 and 2 (page 122), that visual displays quickly reveal information about data and that certain graphs are useful for comparing the frequency of data in different categories.

A glossary is provided at the end of the book, should the reader encounter a mathematical term with which he or she is not familiar. It will also serve as a handy reference for ongoing mathematics teaching.

Although the focus of this resource is on helping teachers understand the mathematics and teach through big ideas, the resource also reflects many other important instructional considerations, including

- the use of effective, and multiple, representations for mathematical concepts;
- the role of manipulatives in making math meaningful to students;
- the value of multiple solutions;
- the value and the potential range of personal strategies and invented algorithms; and
- the importance of teaching conceptually.

Chapter 1

Patterns and Algebra

Patterns

Students study patterns in the primary years for a number of reasons. In part, it helps them understand their everyday surroundings, where patterns are often evident. For example, students might look at patterns in the arrangements of desks or tables in classrooms, shape patterns in the arrangement of tiles on the floor, or number patterns in the street address numbers on a street.

It also supports the rest of their learning in mathematics; students more easily understand many ideas in number, geometry, and even measurement and data when they see the underlying patterns. Students also study patterns because the type of thinking they use to recognize, describe, compare, extend, translate, and create patterns is the basis for the algebraic thinking that will help them succeed in later mathematics.

BIG IDEAS FOR PATTERNS

1. Patterns represent identified regularities. There is always an element of repetition, whether the same items repeat, or whether a "transformation," for example, adding 1, repeats. (BIP 1)
2. The mathematical structure of a pattern can be represented in a variety of ways. (BIP 2)
3. Some ways of displaying data highlight patterns. (BIP 3)
4. Many geometric attributes (e.g., the number of corners or number of faces of various types of prisms), measurements, and calculations involving numbers are simplified by using patterns. (BIP 4)

Teaching Idea | **1.1**

Students can look in books or on the Internet for patterns in nature, for example,

To bring out BIP 1, ask: *What makes the snakeskin design a pattern?* [The brown diamonds are in rows of 2 with a row of 1 between them that's a bit lower and this repeats as you go around the snakeskin.]

A knowledge of big ideas can help teachers choose, shape, and create tasks, and use questioning to help students make powerful connections. Each teaching idea in this section of the chapter will indicate which Big Idea(s) for Patterns (BIP) can be emphasized.

Relating Sorting and Classifying to Patterns

Before they recognize, describe, or extend patterns, students need to sort and classify. We often need to classify or identify objects before we can see a pattern. For example, for students to recognize the repeating pattern at the top of the next page, they need to recognize that there are two different shapes in the pattern. They need to attend to one of the geometric attributes of each shape, number of sides, to differentiate between the two shapes in the pattern.

Sorting and classifying are discussed in Chapter 5 ***pages 115** to **120**.*

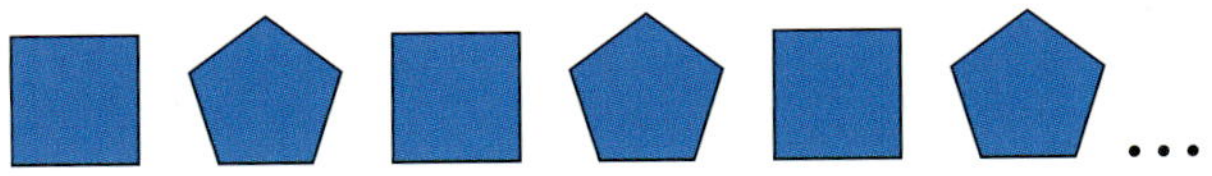

Students need to realize that there is a difference between a square and a pentagon to recognize this pattern.

Types of Patterns

Repeating Patterns

Students' first experiences with patterns generally involve repeating patterns. Initially, the repeating patterns are intuitive. Students hear rhythmic or temporal patterns in sounds, songs, and stories. They recognize patterns in physical actions such as the hokey-pokey game. They also observe patterns in their environment. Students use these patterns to help them predict. Soon after, they are introduced explicitly to repeating colour, shape, size, and position patterns, and somewhat later to repeating number patterns.

The Core of a Repeating Pattern

The shortest part of the pattern that repeats is called the core. In the pattern below, the core is the set of shapes found at the start of the pattern, in this case three shapes, since this is the shortest string that repeats.

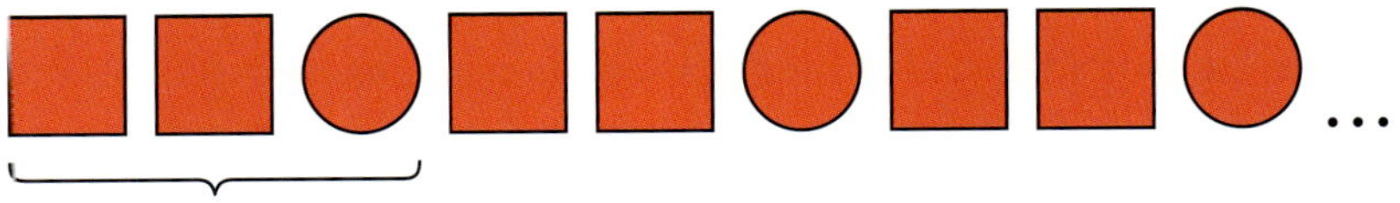

A simple three-element core

It is sometimes difficult for students to determine the core. For example, the following pattern has a five-element core. Students might see the small circle repeating in the fourth spot and guess that the core is one small circle and two large circles. But when they check farther along in the pattern, they will realize that this is not correct and they may look for the next small circle in the pattern to see if the core ends there.

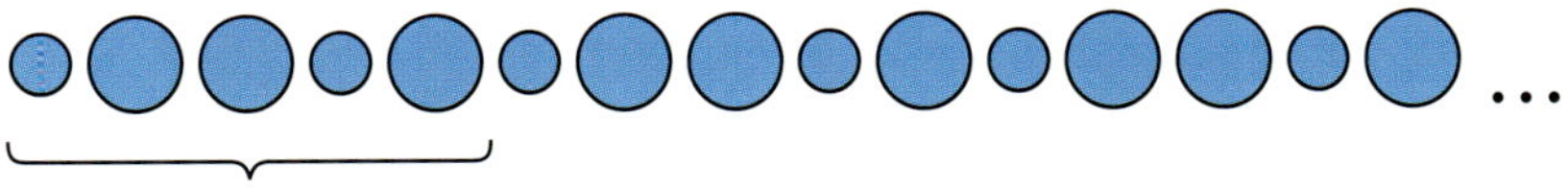

A more complex core

Describing a Repeating Pattern

When students are ready, it is useful to introduce a letter code system to describe repeating patterns. An AB pattern has a core of two different elements that repeat over and over, such as 1, 2, 1, 2, 1, 2, ... An AABC pattern has a core of four elements, where the first two are the same and the others are different, such as 3, 3, 4, 5, 3, 3, 4, 5, 3, 3, 4, 5, ... It is much simpler for students to use a letter code than to record the repetitions and remember to use the ellipsis notation (...) to show that the pattern continues.

Some teachers might like to vary the letter coding, for example, they might call it an XY pattern instead of an AB pattern, to emphasize the arbitrariness of the coding, but other teachers might find it simpler to stick to using the first letters of the alphabet to describe the repetition in a pattern.

Teaching Idea 1.2

Place coloured beads in a row to make a pattern, repeating the core (e.g., blue-red) four or five times. Cover the beads with a paper tent and reveal some of the pattern as shown below. Ask students to predict what they will see next as you slide the tent over.

To focus on BIP 1, ask: *Why did I need to tell you it was a pattern for you to predict the next bead?* [If it were not a pattern, the next bead could be any colour at all.] *Can you be sure what comes next? Why not?* [No; I don't know whether the pattern repeats blue-red or blue-red-blue.]

Teaching Idea 1.3

Tell students that you are thinking about an ABCA shape pattern. Ask them to use pattern blocks to show what your pattern might look like [e.g., red block-yellow block-blue block-red block repeating].

To bring out BIP 2, ask: *How are all of our patterns alike? How are they different?*

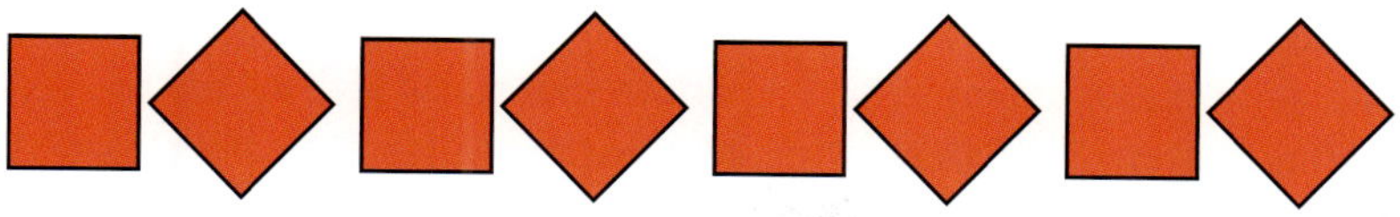

An AB position pattern

An ABA size pattern

The letter code system helps students see that two seemingly dissimilar patterns can be the same mathematically. For example, although all AB patterns are the same mathematically, students tend to focus on the specific elements. Younger students view a pattern made up of a repeating circle and square as different from a repeating number pattern like 1, 2, 1, 2, 1, 2, ... because they focus on the specific elements rather than on the underlying pattern. They may fail to see that the core of all AB patterns is the same, that is, two different items.

The letter code system helps students see that two seemingly dissimilar patterns can be the same mathematically.

Although students in the primary years are sometimes exposed to patterns involving two attributes, for example, size and shape, it is often difficult for students to describe those patterns unless the attributes go together. For example, primary students can understand that the first pattern below shows both colour and shape in an AB pattern. It is much more difficult for them, and probably should not be expected of most of them, to fully describe the second pattern (which is AAB in colour, but AB in shape).

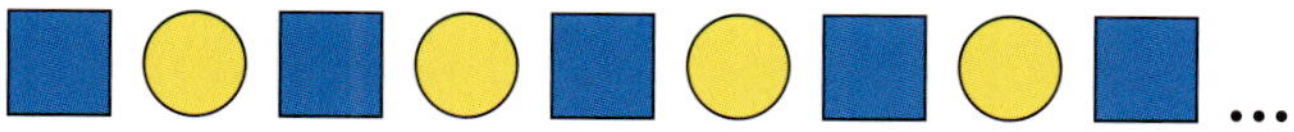

A simple multiple-attribute pattern using shape and colour.
Both shape and colour follow an AB pattern.

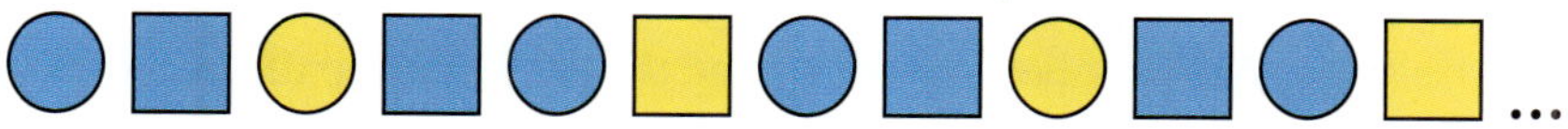

A more complex multiple-attribute pattern using shape and colour.
The shape pattern is AB while the colour pattern is AAB.

Highlighting Repeating Patterns

It is good practice to show at least three full repetitions of the core of a pattern to help students recognize the core. This removes some of the inherent ambiguity. For example, if you just show 2, 4, 6, … of the pattern 2, 4, 6, 2, 4, 6, 2, 4, 6, …, students might not be sure whether the pattern repeats or whether it grows in a +2 pattern. Even if we show 2, 4, 2, 4, …, students might suspect that 2, 4 comes next, but they will feel more confident if there is at least one more repetition of the core. However, even with three repetitions, you can never be certain about a pattern. For example, 2, 4, 6, 8, … could be 2, 4, 6, 8, 2, 4, 6, 8, …, or 2, 4, 6, 8, 6, 4, 2, or 2, 4, 6, 8, 10, 12, ….

It is important that students understand that, even with three repetitions of the core of a pattern, you can never be certain about a pattern.

Teaching Idea 1.4

Show students an ABC colour pattern with 12 elements, using counters or square tiles (e.g., blue-yellow-red repeating). Then rearrange the counters or tiles into two different arrangements, as shown below.

To bring out BIP 3, ask: *Which way of showing the first 12 shapes in the pattern makes it easier for you to see what comes next?* [A] *Why?* [In A, each row is one repeat of the pattern core, so it's easy to predict the next row.]

Arrangement A

Arrangement B

In higher grades, students learn about formal pattern rules that define a unique pattern, for example, "a +2 pattern that starts with 4," or "an AABA repeating pattern." As they work with pattern rules, they will learn that, without a pattern rule, one can never be sure how a pattern continues. It makes sense, at the K to 3 level, to plant that seed by occasionally asking, after students suggest an extension to a pattern, whether they can be sure the pattern continues in the suggested way or whether there is another possible extension.

The way a pattern is arranged often makes it easier or more difficult for students to recognize and describe it. For example, you might separate the elements that make up the core for very young students, as shown below.

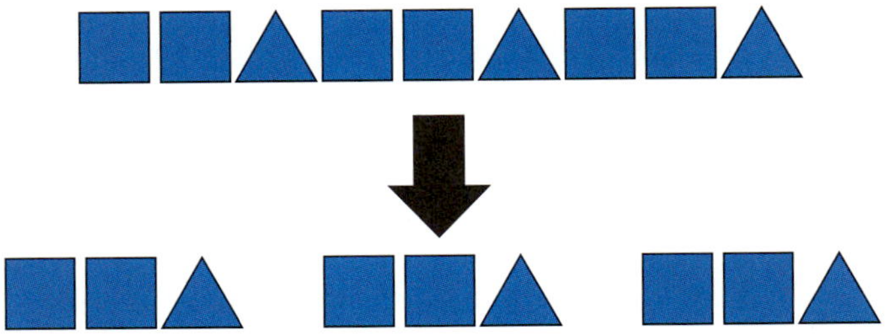

Separating the elements that make up the core to highlight a repeating pattern

Growing and Shrinking Patterns

In the primary years, students work with very simple growing and shrinking number or shape patterns. For number patterns, a small constant value, often 1 or 2, is added to or subtracted from each value or number of shapes to get the next value or number of shapes.

1, 2, 3, 4, …   2, 4, 6, 8, …   10, 8, 6, …

Occasionally, students might meet a pattern that includes both repetition and growing or shrinking, although this is more likely in higher grades than in the K to 3 range. For example, the pattern 1, 2, 1, 3, 1, 4, 1, 5, … involves repetition (all of the odd terms are the same), but it also involves growing (the even terms go up by 1 each time).

Representing Growing and Shrinking Patterns

For the shape patterns below, notice that the first shape pattern grows in terms of the number of items in each figure, whereas the second shape pattern grows in terms of the number of sides of each figure or shape.

The green shape pattern highlights BIP 4: Many geometric attributes … are simplified by using patterns.

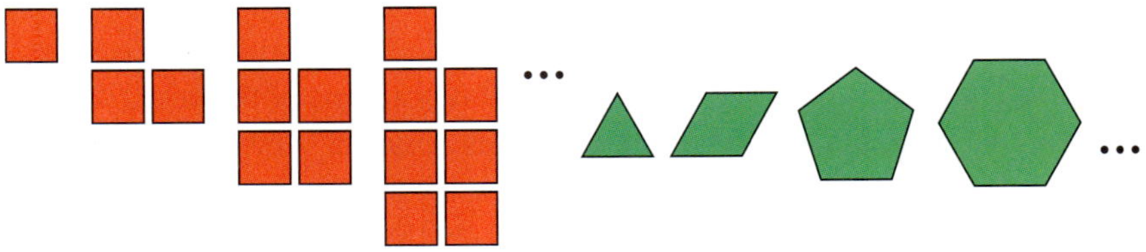

Other growing and shrinking patterns use numbers in a less obvious way because they focus on measurement. For example, the Cuisenaire rod pattern at the top of the next page is a simple shrinking pattern because the rods get shorter by 1 cm each time.

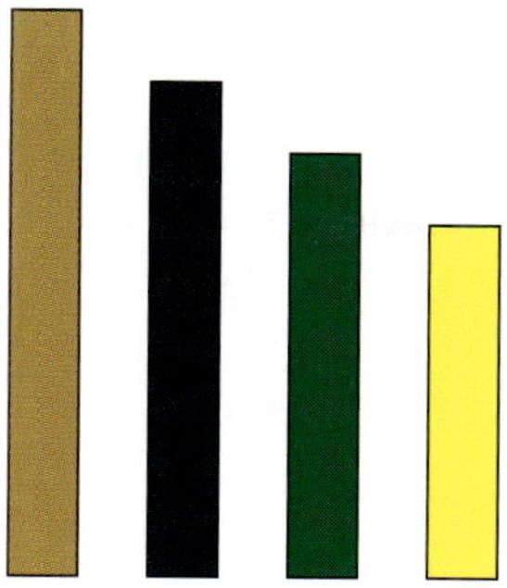

A shrinking measurement pattern made with Cuisenaire rods

Highlighting Growing and Shrinking Patterns

As with a repeating pattern, it is good practice to show at least three repetitions of the change in a growing or shrinking pattern. For example, a pattern such as 3, 6, 9, … appears to be a growing pattern but it could also be the repeating pattern 3, 6, 9, 3, 6, 9, 3, 6, 9, …. Including just one more repeat of the +3 (3, 6, 9, 12, ...) aspect of the pattern will give students more confidence to extend the pattern.

You can also arrange elements in a pattern to highlight growing and shrinking patterns. In the example below, the number pattern in the red squares (1, 3, 5, 7, …) is highlighted when the squares are arranged to emphasize the fact that two squares are added each time; this makes it easier for students to identify the pattern.

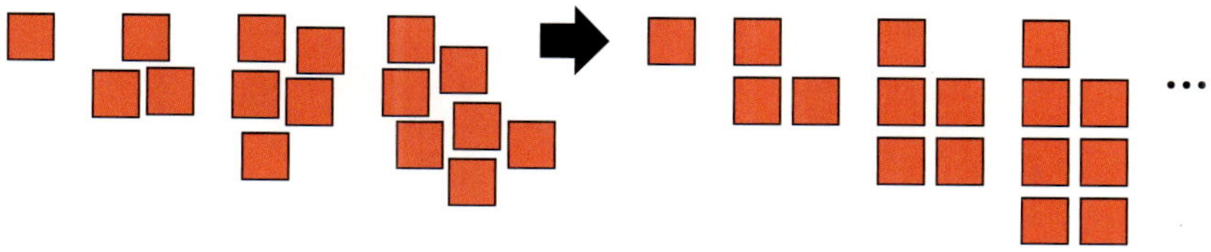

Rearranging the elements in a growing pattern to make the pattern more obvious

Describing and Extending Patterns

To describe a repeating pattern, students might use a letter code (e.g., ABA) or they might state the repeating core.

Students seem to find it easier to extend a repeating pattern that ends with a full repetition of the core than to extend a pattern that ends in the middle of the core. For example, students might have difficulty extending the pattern *circle*, *circle*, *square*, *circle*, *circle*, *square*, *circle*, *circle*, *square*, *circle*, …. They often repeat the full core again, ignoring the fact that part of core was already included. Students find it much easier if the pattern is stated as *circle*, *circle*, *square*, *circle*, *circle*, *square*, *circle*, *circle*, *square*, ….

To describe growing or shrinking patterns, primary students usually use informal pattern rules. Rather than expecting the precision of a formal rule, where they state the first term and indicate the repetition, allow students to describe the pattern loosely. For example, for 2, 4, 6, …, they are likely to assume that the start of 2 is understood and simply say that it is an "add 2" pattern, rather than expressing the complete pattern rule (start with 2 and add 2 to get to the next value).

Teaching Idea 1.5

Show students a pattern such as the measurement pattern shown to the left.

To bring out BIP 4, ask: *Describe another pattern that uses measurements that grow bigger or smaller in a pattern.* [e.g., Four of the same glasses in a row with water in them. Each glass contains three more big spoonfuls of water than the glass to its left.]

Teaching Idea 1.6

Provide students with 30 linking cubes, all the same colour. Ask them to use the cubes to show enough of a growing pattern so that another student could figure out the pattern, for example, show 1, 3, 5, 7, 9, … cubes.

To bring out BIP 1, ask: *What will your partner say repeats in your growing pattern?* [Adding 2 repeats.]

Teaching Idea 1.7

Ask a group of seven students to line up and act out a repeating pattern with this core: face front-face back.

To bring out BIP 1 and BIP 3, ask: *I am going to invite two more students to come up. Which way should they face to continue the pattern?* [face front, then face back] *Can you be sure?* [e.g., I can't be sure because you didn't tell us what the core of the pattern was, but it looks right.]

Young students are likely to assume that there is only one way to extend a given pattern. The fact that the pattern 2, 3, 2, 3, 2, 3 … could be extended in any of the ways below does not occur to most primary students.

2, 3, 2, 3, 2, 3, 2, 3, 2, 3, 2, 3, …
2, 3, 2, 3, 2, 3, 3, 4, 3, 4, 3, 4, …
2, 3, 2, 3, 2, 3, 1, 2, 3, 2, 3, 2, 3, 1, …

Teaching Idea 1.8

Begin a repeating sound pattern with a core such as *clap-clap-snap-snap-clap*. Tell students that this is the core of a repeating pattern.

To bring out BIP 2, ask: *How could you use square and circle counters to represent the same pattern?* [core: square-square-circle-circle-square] *How else could you represent the pattern using attribute blocks?* [core: circle-circle-square-square-circle]

Translating Patterns

If you want to know whether students are focused on the mathematical structure of a pattern, ask them to translate the pattern into a different but equivalent form. To translate a pattern means to represent the same pattern structure, but with different elements. For example, show students the shape pattern below, marked Original Pattern, and ask them to show the same kind of pattern using other shapes. There are many possibilities, as shown below. The first translated AAB pattern changes by shape, the second by colour, and the third by position and colour. Using a letter code to describe a repeating pattern is also a way of translating a pattern.

Original AAB Pattern

Examples of Translated AAB Patterns

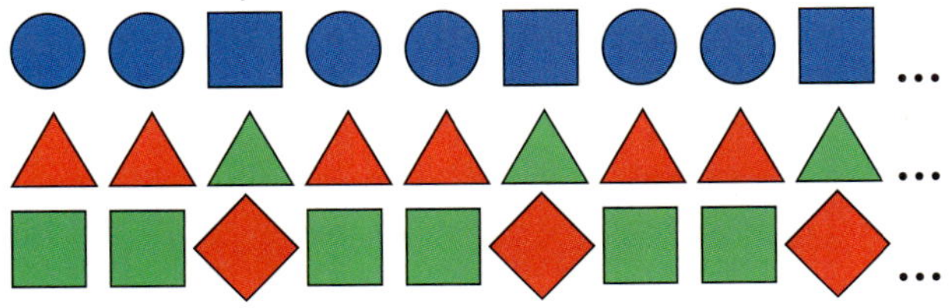

Teaching Idea 1.9

Ask students to create a number pattern that has more even numbers than odd numbers in the first 30 numbers [e.g., 2, 4, 5, 2, 4, 5, 2, 4, 5, …]. Ask: *How do you know there will be more even numbers than odd numbers?* [The core has more even numbers than odd numbers, so the whole pattern will have more evens than odds.]

To bring out BIP 1, ask: *How can you convince someone that it is a pattern?* [e.g., I can make sure he or she notices that the same numbers repeat.]

Creating Patterns

Encourage students to create patterns as soon as they understand what patterns are. As they are exposed to many types of patterns, and patterns of various structures based on many different attributes, whether quantity, shape, orientation, or size, students' patterns will become richer.

Most young students need to create patterns with concrete materials before they make pictorial representations and then move on to number patterns at a more abstract level. They are likely to be more comfortable working with repeating patterns before they explore growing patterns.

Sample Criteria for Creating Patterns

You can use many criteria to structure students' pattern creation. You can impose less or more structure when you tell them what kind of pattern to make. For example, you might ask students to create

- a shape pattern that has triangles in it
- a shape pattern where the fourth shape is a triangle
- a number pattern like the pattern shown in **Teaching Idea 1.9**
- a measurement pattern where lengths or areas get smaller

If you require specific criteria, students will be forced to use more reasoning skills, but struggling students will find it more challenging.

Mathematical Situations Rich in Patterns

A number of charts and tables rich in number patterns are used in mathematics classrooms to build number concepts. (BIP 4: Many … calculations involving numbers are simplified by using patterns.) As students work on tasks like these, make sure they notice that patterns can help them with number concepts. For example, when working with patterns in 100 charts, you might ask, *How does the pattern in the chart help you add 10 to a number*? [Go straight down to the number right below it in the next row.]

You can use number patterns in charts and tables to build number concepts (BIP 4).

Calendar Patterns

The calendar is one of the first places where young students are exposed to number patterns. Some of the many patterns that students might observe using the calendar are described below.

Sunday	Monday	Tuesday	Wednesday	Thursday	Friday	Saturday
	1	2	3	4	5	6
7	8	9	10	11	12	13
14	15	16	17	18	19	20
21	22	23	24	25	26	27
28	29	30				

Teaching Idea 1.10

To focus on BIP 3, ask: *How does showing the days of a month using a 7-day calendar pattern help you see that September 26 is exactly 2 weeks after September 12?* [Two weeks is 2 sets of 7 days, or 2 rows, so September 26 is two rows down in the same column.]

COLUMN PATTERNS

You always add 7 when you go down a column.

6
13
20
27

MONTHS OF THE YEAR PATTERN

The pattern of the months repeats as you go through years. January, February, March, April, May, June, July, August, September, October, November, December is the core of a 12-element repeating pattern of months, year after year.

DAYS OF THE WEEK PATTERN

The days of the week form a 7-element repeating pattern with a core of Sunday, Monday, Tuesday, Wednesday, Thursday, Friday, Saturday.

ROW PATTERNS

You always add 1 when you go across a row.

7	8	9	10	11	12	13

DIAGONAL PATTERNS DOWN AND TO THE RIGHT

Each arrow like the red one adds 8 to the first number. This is because you are going down a row, which adds 7, and one space to the right, which adds 1 (and 7 + 1 = 8).

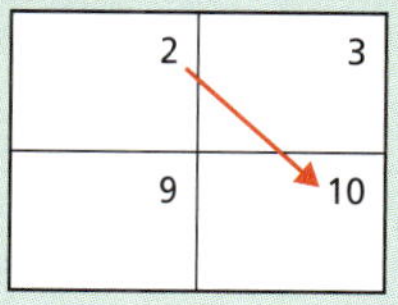

DIAGONAL PATTERNS DOWN AND TO THE LEFT

Each arrow like the blue one adds 6 to the first number. This is because you are going down a row, which adds 7, and one space to the left, which subtracts 1 (and 7 − 1 = 6).

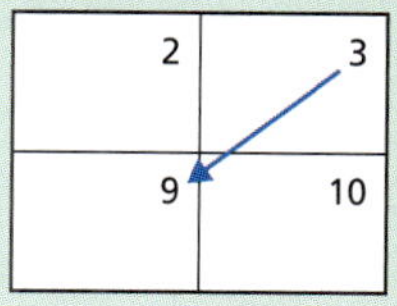

10-Frame Patterns

10-frames are also discussed in Chapter 2 on ***pages 22*** *to* ***23****.*

Early in their work with numbers, students meet the 10-frame, which is a 2 row-by-5 column grid. When they work with a 10-frame, students generally place a counter (or shade a square) to the right each time the number that is being represented increases by 1 until the 5 squares in the top row are shaded. Then they place counters or shade squares in the next row until the frame is full. They may use additional 10-frames if they need to display a larger quantity. The visual patterns emphasize the numerical patterns that they model, as shown below.

Example 1 Pattern: 1, 2, 3, 4, 5, ...

These 10-frames show that the numbers are increasing by 1.

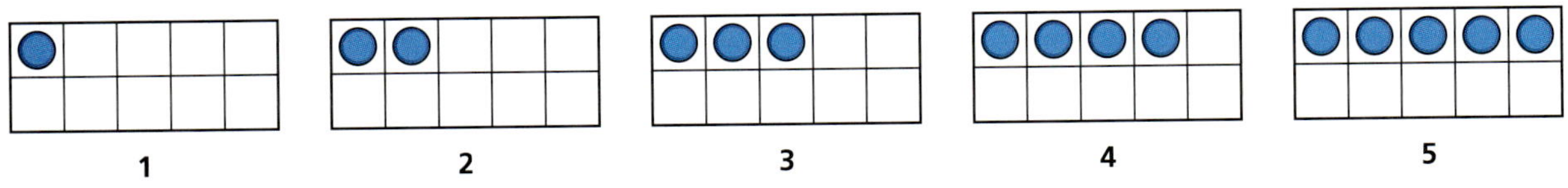

Example 2 Pattern: 5, 10, 15, 20, ...

These 10-frames show that the numbers increase by 5 because another full row of 5 is filled each time.

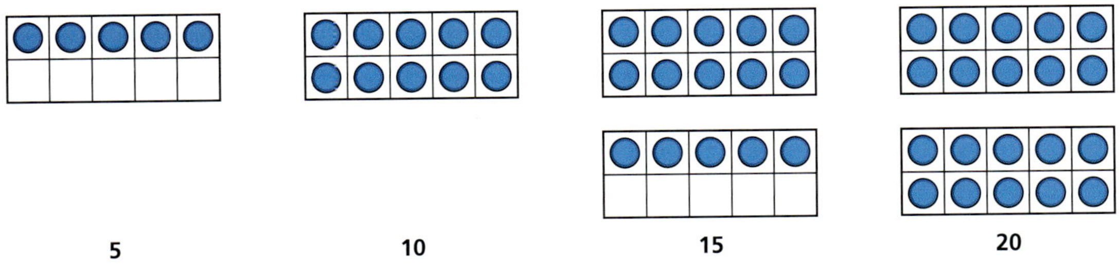

Example 3 Pattern: 4, 14, 24, 34, ...

These 10-frames show that the numbers increase by 10 because a full 10-frame is added each time. This explains why adding 10 does not affect the ones digit.

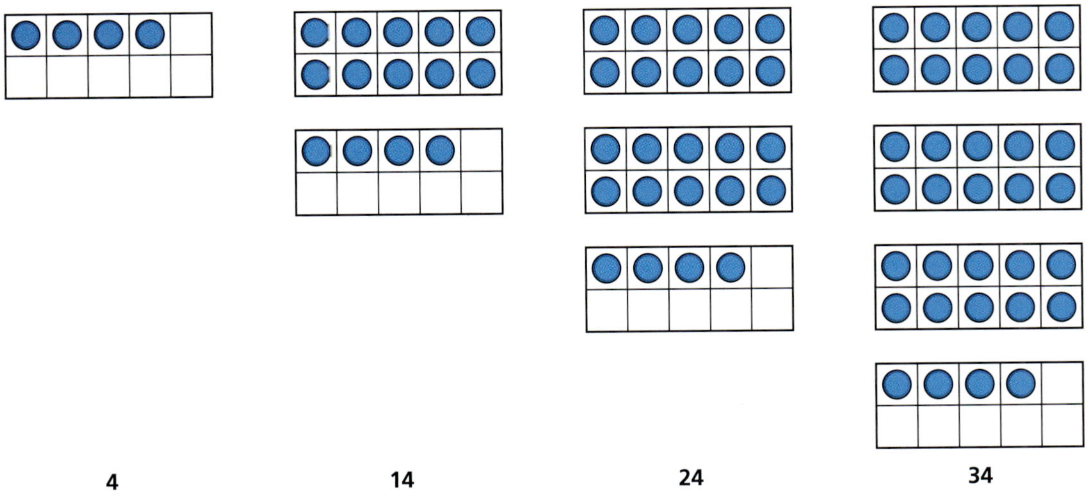

100 Chart Patterns

The 100 chart is a tool students can use through elementary and middle school grades. Some of the patterns in the 100 chart are shown on the next page.

100 charts are always consecutive numbers shown in order in 10 rows. Some teachers use a chart that begins at 0 and goes to 99, so that all of the numbers in any row have the same tens digit. Others use the numbers from 1 to 100 because students more naturally begin with 1 than 0 when they count and because they want to emphasize the idea that the number 100 is 10 tens. Using a Smart Board, teachers might easily create such charts or even switch from one type to the other.

In the examples shown below, the latter type of chart is used, but the ideas could be easily adapted if a 0 to 99 chart is used.

Many of the patterns that appear in the 100 chart can be extended if charts showing the numbers from 101 to 200 or even to 300 or 400 are created.

PLACE VALUE PATTERNS IN 100 CHART

- Every 10th number ends in a 0.
- The pattern of ones digits is always 1, 2, 3, 4, 5, 6, 7, 8, 9, 0, repeating over and over.
- The tens digits of the 2-digit numbers form groups of 10 numbers that repeat and then jump by 1:
  1, 1, 1, 1, 1, 1, 1, 1, 1, 1;
  2, 2, 2, 2, 2, 2, 2, 2, 2, 2;
  3, 3, 3, 3, 3, 3, 3, 3, 3, 3;
  4, 4, 4, 4, 4, 4, 4, 4, 4, 4; ....
- If you go down a column, you keep adding 10.
- If you go across a row to the right, you keep adding 1.

1	2	3	4	5	6	7	8	9	10
11	12	13	14	15	16	17	18	19	20
21	22	23	24	25	26	27	28	29	30
31	32	33	34	35	36	37	38	39	40
41	42	43	44	45	46	47	48	49	50
51	52	53	54	55	56	57	58	59	60
61	62	63	64	65	66	67	68	69	70
71	72	73	74	75	76	77	78	79	80
81	82	83	84	85	86	87	88	89	90
91	92	93	94	95	96	97	98	99	100

Teaching Idea **1.11**

After students have explored patterns in a 100 chart, ask them to visualize what the chart would look like if it had 5 columns instead of 10.

To bring out BIP 1 and BIP 3, ask: *Which patterns in the chart would stay the same?* [e.g., All numbers that end in 0 would be in the same column.] *Which patterns would be different?* [e.g., Even numbers would not all be in the same columns.] *Would there be new patterns?* [e.g., Even numbers would form a checkerboard pattern.]

Addition Table Patterns

The addition table is a way for students to summarize information about the facts they have learned. It is also a tool they can use to learn new facts based on patterns. If some numbers are included on the chart and other numbers are omitted, students can use the chart to figure out the missing sums.

Row and Column Patterns

- The numbers go up by 1 as you go down columns or across rows.
- The numbers alternate between even and odd values.
- The number of two-digit numbers increases by 1 as you move to the next row or the next column (10, then 10, 11, then 10, 11, 12, and so on).

Diagonal Patterns

- On any diagonal going down from left to right, the numbers increase in a + 2 pattern.
- On any diagonal going up from left to right, the numbers are the same.

Square Patterns

In any 2-by-2 square,

- the top right and bottom left numbers are equal (e.g., 12 = 12)
- the top left number is 2 less than the bottom right number (e.g., 13 − 2 = 11)
- the sums of the numbers on the diagonals are equal (e.g., 11 + 13 = 12 + 12)

11	12
12	13

+	0	1	2	3	4	5	6	7	8	9
0	0	1	2	3	4	5	6	7	8	9
1	1	2	3	4	5	6	7	8	9	10
2	2	3	4	5	6	7	8	9	10	11
3	3	4	5	6	7	8	9	10	11	12
4	4	5	6	7	8	9	10	11	12	13
5	5	6	7	8	9	10	11	12	13	14
6	6	7	8	9	10	11	12	13	14	15
7	7	8	9	10	11	12	13	14	15	16
8	8	9	10	11	12	13	14	15	16	17
9	9	10	11	12	13	14	15	16	17	18

Multiplication Table Patterns

The multiplication table is a way for students to summarize information about the facts they have learned. It is also a tool they can use to learn new facts based on patterns. If some numbers are included on the chart and other numbers are omitted, students can use the chart to help them figure out the missing products.

Row and Column Patterns

- The numbers in row *x* match the numbers in column *x* (e.g., the numbers in the 4 row match the numbers in the 4 column).
- The numbers increase by the row number as you go down columns or across rows (e.g., the numbers in the row for 5 increase by 5).
- Every other row or column is made up only of even numbers. The other rows and columns alternate between even and odd.
- When you add the matching values for two rows or columns (if the sum for those rows or columns is less than 10), the sums are in another row or column on the chart (e.g., if you add the matching values for the 3 row and the 6 row, the sums are in the 9 row since $3 + 6 = 9$).

3	0	3	6	9	12	15	18	21	24	27
6	0	6	12	18	24	30	36	42	48	54
9	0	9	18	27	36	45	54	63	72	81

The 3 row + the 6 row = the 9 row

Square Patterns

In any 2-by-2 square, the sum of the top right and bottom left numbers is one less than the sum of the numbers on the other diagonal.

12	15
16	20

For example, $12 + 20 = (15 + 16) + 1$.

The product of the pairs of numbers on the two diagonals are equal. For example,

$4 \times 10 = 8 \times 5$.

4	5
8	10

X	**0**	**1**	**2**	**3**	**4**	**5**	**6**	**7**	**8**	**9**
0	0	0	0	0	0	0	0	0	0	0
1	0	1	2	3	4	5	6	7	8	9
2	0	2	4	6	8	10	12	14	16	18
3	0	3	6	9	12	15	18	21	24	27
4	0	4	8	12	16	20	24	28	32	36
5	0	5	10	15	20	25	30	35	40	45
6	0	6	12	18	24	30	36	42	48	54
7	0	7	14	21	28	35	42	49	56	63
8	0	8	16	24	32	40	48	56	64	72
9	0	9	18	27	36	45	54	63	72	81

Algebra

Most people think that algebra is math with letters. However, mathematicians view algebra as a system that allows them to

- represent mathematical relationships
- explain relationships among quantities
- analyze change

BIG IDEAS FOR ALGEBRA

1. Algebra is a way to represent and explain mathematical relationships and to describe and analyze change. (BIA 1)
2. Relationships between quantities can be described using variables. (BIA 2)

Each teaching idea in this section of the chapter will indicate which Big Idea(s) for Algebra (BIA) can be emphasized.

Algebra and Young Learners

Virtually all of the algebraic work in Kindergarten to Grade 3 is related to number relationships and change.

Teaching Idea 1.12

Ask students to write addition fact families that contain 5 as one of the numbers [$1 + 5 = 6$, $5 + 1 = 6$, $6 - 1 = 5$, $6 - 5 = 1$; $2 + 5 = 7$, $5 + 2 = 7$, $7 - 2 = 5$, $7 - 5 = 2$, and so on]. Then ask: *How many sentences were in each fact family?* [Usually four, but one family (the 5, 5, 10 family) had two.]

To focus on BIA 1, ask: *Can you predict other fact families that have only two sentences?* [Yes, any family that has a double addition fact, like $4 + 4 = 8$, $6 + 6 = 12$.]

GENERALIZING CHANGE WHEN ADDING 1

As young students observe that adding 1 means saying the next number in the counting sequence, possibly using a walk-on number line, they are representing a mathematical relationship that is essentially algebraic.

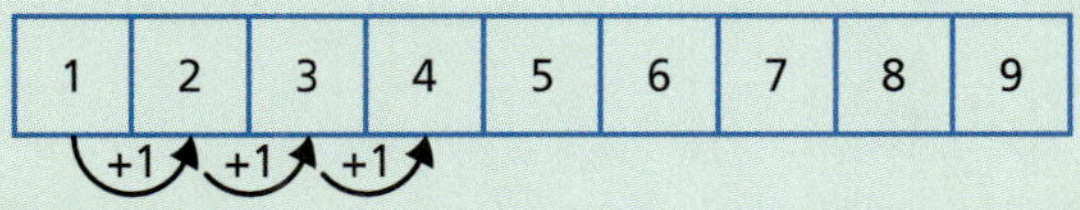

GENERALIZING CHANGE WHEN ADDING 2

As they write equations like $4 + 2 = 6$, they are representing a specific relationship and change, the relationship between 4 and 6 and the change required to change 4 to 6. When they realize that the addition of 2 always results in a number that is the second one after the given number in the counting sequence, they have come to an algebraic generalization.

The earlier section in this chapter on patterns spoke to how young students build capacity for algebraic thinking by studying patterns. Young students also develop and use algebraic thinking when working in the number strand, when they work with simple open sentences, fact families, and simple equations. For example,

- when students solve $3 + \square = 8$ by thinking "if $3 + \square = 8$, then $\square = 8 - 3$," they are applying the general algebraic relationship between addition and subtraction: if $a + b = c$, then $c - a = b$.
- when they write the fact family for 3, 4, and 7 ($3 + 4 = 7$, $4 + 3 = 7$, $7 - 3 = 4$, and $7 - 4 = 3$), they are applying the general algebraic relationship between addition and subtraction: if $a + b = c$, then $b + a = c$, $c - a = b$, and $c - b = a$.

- when they compare the sums of 3 + 5, 4 + 5, 5 + 5, ... they are analyzing change; as you add 1 more each time to one of two numbers being added, the sums increase by 1.

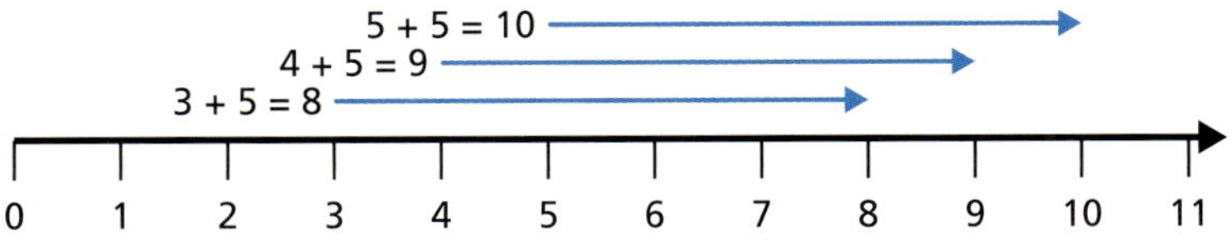

Students can generalize that, if $a + b = c$, then $(a + 1) + b = c + 1$.

Using Variables

In the primary years, most students are not comfortable using letter variables to represent unknowns. Initially, it is helpful to use open boxes as variables. For example, in the equation 3 + ❑ = 7, the ❑ is an unknown, or variable. It might also be advisable to use other shaped open boxes to help students see that the actual shape of the box used is irrelevant; 3 + ❑ = 7 could also be written as 3 + △ = 7 or 3 + ○ = 7. The open box suggests intuitively that something is missing. Later on, as a transition to letter variables, which are used in the later grades, a symbol such a solid black square, circle, or triangle could be used, for example, 3 + ❑ becomes 3 + ■.

If you are using more than one unknown in a single situation, it is best to use the same symbol if the unknown has the same value, and a different symbol if it might have a different value. For example, you might write 3 + ■ = ▲ to show that the two unknowns have, or could have, different values, but you would write 3 + ■ = ■ + 3 to show that you can add any number to 3 in either order.

Teaching Idea 1.13

Ask students to use the following symbols and numbers to write three equations:

4, 5, 7, 8, ■, ■, ■, ▲, ▲

[e.g., 5 + ■ = 7; ■ + 4 = ▲; and ■ + ▲ = 8]

To bring out BIA 2, ask: *What does the first number sentence tell you about how 5 and 7 are related?* [5 is a certain amount less than 7, and 7 is a certain amount more than 5.] *What do the other two sentences tell about the unknown numbers?* [The second sentence tells that ■ and ▲ are 4 apart and that ▲ is greater. The third sentence tells that ■ and ▲ add to 8.]

Exploring Relationships

As stated earlier, most of the relationships that primary students explore that are algebraic in nature relate to numbers. For example, as students look at the effect of subtracting 3, they begin to make the general observation that the numbers decrease. Eventually they will see that the difference is 3 less than the number they started with.

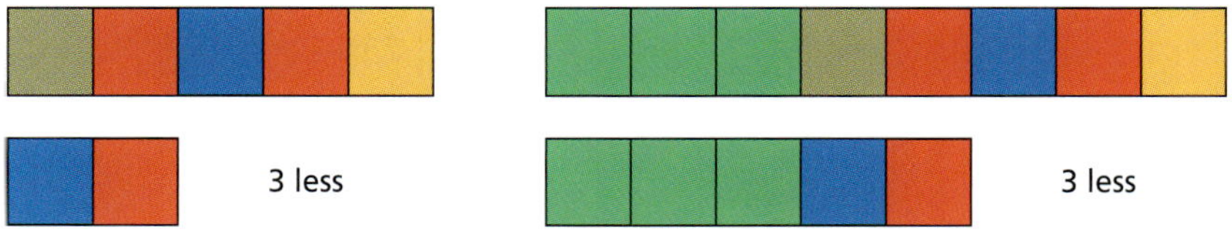

5 − 3 = 2 and 2 is 3 less than 5   8 − 3 = 5 and 5 is 3 less than 8.

Students need to generalize that subtracting 3 results in a value that is 3 less than the number they started with.

Here are some examples of other relationships that primary students might use:

- Adding 9 by adding 10 and then subtracting 1 is using the general relationship that since 9 is 1 less than 10, increasing something by 9 is the same as increasing it by 10 and then decreasing it by 1.
- The place value notion of adding 10 by simply increasing the tens digit of a number by 1, for example, <u>2</u>4 + 10 = <u>3</u>4, is based on the general principle that you can regroup and add numbers in different orders, 20 + 4 + 10 is the same as (20 + 10) + 4.

- Students might observe the generalization that regrouping 10 tens for 1 hundred or 10 ones for 1 ten decreases the total number of base ten blocks required to model a number by 9. For example,

12 can be represented by 12 ones, which is 12 blocks, or by 1 ten and 2 ones, which is only 3 blocks.

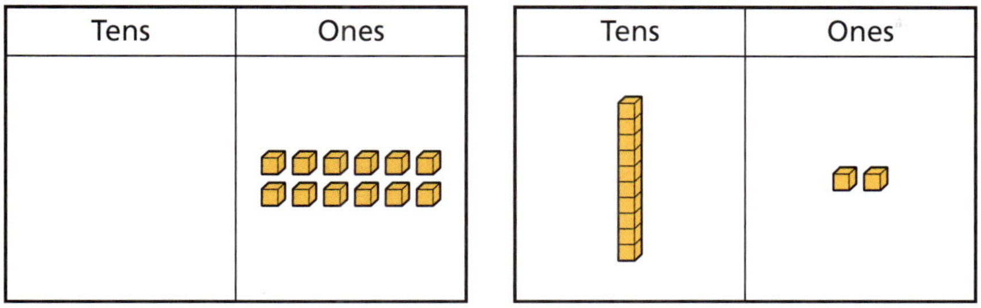

Students might generalize that, when you regroup 10 ones for 1 ten, you use 9 fewer blocks.

- Multiplying by 8 by multiplying by 4 and then doubling (e.g., 8×6 is double 4×6) since 8 groups of something is twice as much as 4 groups of that something is a general algebraic principle.

Exploring Equality

Primary students should work concretely with the metaphor of an equation as a balance starting in or before Grade 1. For example,

Students know that $4 + 8 = 12$, since 12 cubes on one pan of a balance scale can balance 4 and 8 on the other pan.

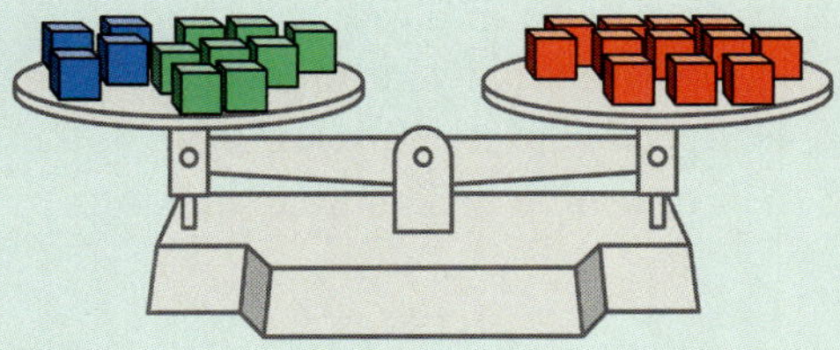

They also know that $\blacksquare = 8$ in $4 + \blacksquare = 12$, since if they have 12 cubes on one pan and 4 on the other pan, they must add 8 cubes to the 4 to balance the scale.

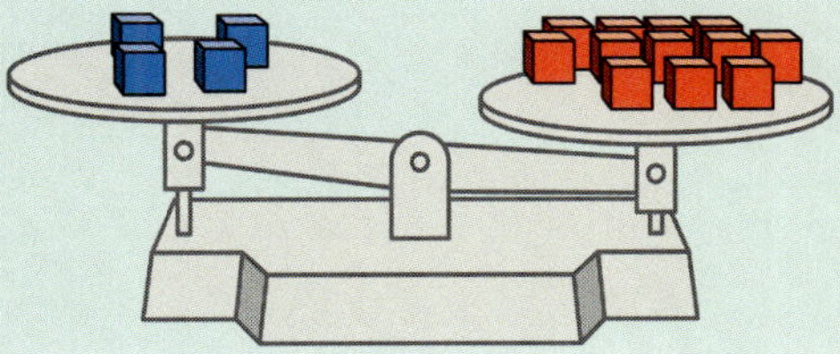

Primary students should explore equations where the unknown is sometimes on the left and sometimes on the right to help them understand that the equals sign describes a balance; the equals sign is not an instruction to perform a particular operation, a common misconception that is reinforced by the use of the equals key on a calculator. Many students without this knowledge solve $4 + \blacksquare = 12$ by adding 4 and 12 since they see a 4, a +, and a 12.

Teaching Idea 1.14

Set up a pan balance as shown below, with 3 cubes in a light paper bag to represent the unknown, or variable. (Ensure students know that the bag's weight does not significantly affect the balance.)

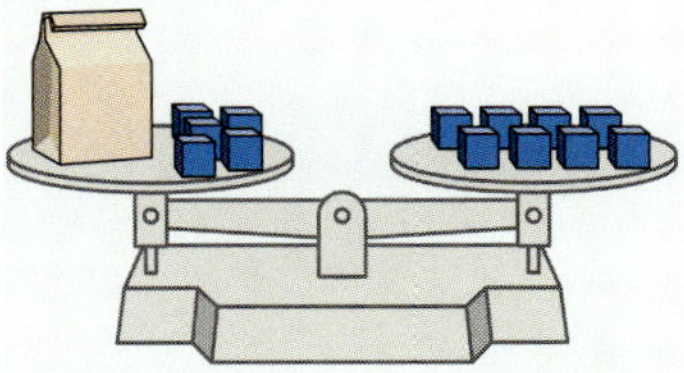

To focus on BIA 2, ask: *What equation describes what you see on the scale?* [e.g., $\blacksquare + 5 = 8$]

To focus on BIA 1, ask: *How would the equation change if I put one more cube on each side of the scale?* [e.g., $\blacksquare + 6 = 9$] *How do you know the pans would stay balanced?* [The pans were balanced to start with and you put the same amount on each side.]

It is important that students deal with equations involving both addition and subtraction and, once they have been introduced to multiplication and division equations involving those operations as well, for example, $3 \times 4 = ■$ or $12 \div ■ = 4$. It is also advisable for students to meet some equations involving two operations, for example, both addition and subtraction are used in $3 + ■ - 1 = 8$.

Teaching Idea | **1.15**

Tell students that you are thinking of a secret number less than 8, and you are going to add 1 to your secret number.

To focus on BIA 1, ask: *What do you know for certain about the new value?* [It is less than 9.]

Exploring Inequality

Although we do not introduce the inequality signs (< and >) formally to young students, it is possible to do so. Whether or not you choose to use these signs, it is important to have students compare expressions to decide whether one is greater or less than another. This is an important part of building number sense. For example, you might ask students which is greater in each pair: $3 + 2$ or $3 + 3$, $10 + 2$ or $10 + 3$, and $18 + 2$ or $18 + 3$. Although exploring each individual comparison is not, in itself, algebra, consideration of concepts such as the fact that $3 + a$ is always less than $3 + (a + 1)$ is an algebraic relationship.

Exploring Change

As students work with number, it is important to consider the underlying algebra of change.

Much of mathematics is built around transformations, or the notion of change. For example, we consider what "adding 3" does to a number or what "multiplying by 1" does to a number. As students work with number, it is important to consider the underlying algebra of change. For example, when you ask students what is $324 + 100$, they are doing number work. When you ask them to generalize what happens to any number when they add 100, you are asking them to think algebraically. In this case, you would expect them to talk about how the hundreds digit increases by 1; this is a generalized description of change.

The notion of change is also important in geometry and measurement, but the relevant concepts are usually introduced at a higher grade level. The more a primary teacher can help students see some of these generalizations, the easier will be their transition to algebraic thinking.

Chapter 2

Number and Operations

Early Number

Numbers are a part of every young child's life. He or she wants to know how many of something there are, whether it is how many cookies he or she can eat, how many fingers someone is holding up, or how many candles are on his or her cake. Before they can work with numbers, whether representing quantities or performing operations, children learn to count.

A knowledge of big ideas can help teachers choose, shape, and create tasks, and use questioning to help students make powerful connections.

BIG IDEAS FOR EARLY NUMBER

1. A number tells how many are in a group. You usually count to determine the size of a group. (BIEN 1)
2. Counting is fundamental to number. Forms of counting include rote counting, counting all, counting on, skip counting, and counting back. (BIEN 2)
3. You can represent a number in a variety of ways. Each representation of a number can focus on a different aspect of the number. (BIEN 3)
4. To compare the numbers of items in two sets, you can match the items, one to one, in the two sets to see whether one set has more. Or, you can compare the position of the numbers that describe the two quantities in the number sequence. (BIEN 4)
5. Students gain a sense of the size of numbers by comparing them to meaningful benchmark numbers. (BIEN 5)

Each teaching idea in this section of the chapter will indicate which Big Idea(s) for Early Number (BIEN) can be emphasized.

Counting Principles

Although students may recognize some quantities without counting, some form of counting is normally required for values beyond 5. To do this, a first step is that students are able to recite the sequence that begins 1, 2, 3, 4, 5, …. This is called rote counting. How high students can count depends on how much of the sequence they know.

If teachers are faced with students who cannot rote count, they can use songs, rhymes, and other oral practice to familiarize students with the

Teaching Idea **2.1**

Display eight objects and deliberately count them incorrectly. You might say "1, 2, 3, 5, 6, 8, 9, 10. There are ten objects here." Repeat to make sure students are aware you miscounted.

To focus on BIEN 1 and 2, ask: *Do you agree that there are ten objects?* [no] *Why did my counting not work?* [You missed some numbers.] *Can I know how many there are without counting?* [yes, if there are only a few objects, but usually you have to count]

sequence of numbers. It is essential, too, that students learn that each word is separate; some students simply hear a long set of sounds "onetwothree…". Sometimes students will know parts of the sequence, for example, "one two three" and "eight nine ten," but it is, of course, necessary that the entire sequence be put together.

A teacher might provide practice for students in saying the next number in a sequence in a variety of ways. She might sometimes provide all the numbers that come before the missing numbers, and other times only one or a few of them. For example, a teacher could say, *What number comes next?*

One, two, three, four, …   Three, four, …   Five, six, seven, eight, …

Many, but not all, students can rote count to 10 or higher. Many can count a small set of objects meaningfully, in other words, they can correctly associate the appropriate number with a set of objects, before they begin school. If they can, they have begun to internalize the following fundamental counting principles.

COUNTING PRINCIPLES

1. You say one and only one number for each object (the one-to-one principle).
2. There is a consistent set of counting words that never changes (the stable order principle).
3. The last number spoken tells how many (the cardinal principle).
4. It does not matter what you count; the process for counting remains the same (the abstraction principle).
5. It does not matter in which order you count; the number in the set does not change (the order-irrelevance principle).

(Gelman and Gallistel, 1978)

Some students take much longer than other students to become successful counters.

As can be seen by looking at the list of principles, counting is fairly complex for young students. A lot of separate things are going on simultaneously—keeping track of what has to be counted, what has been counted, and the number words that must be said, as well as knowing when to stop. Some students take much longer than other students to become successful counters.

In some jurisdictions, the notion that counting represents forward or backwards movement and the notion that unitizing represents counting in groups of 10 as well as groups of 1 are also considered counting principles. Both of these ideas will be discussed later on in this chapter.

Counting Strategies

Students tend to use a variety of physical techniques to help them count correctly. Some teachers worry about students using their fingers to count, but this is a natural way for some students to keep track. Other students use other strategies for keeping track of what they have counted, such as touching each object or moving it away once it has been counted as they say the counting numbers. Later, they learn to use visual or other kinesthetic motions, like nodding, to keep track of what has been counted. Students normally learn these techniques by copying others who are efficient counters, such as fellow students or the teacher.

The number line can be a visual tool to help students count. Teachers who introduce the number line can help students make sense of it by making the connection between movement and quantity.

For example, as shown below, when moving the 7 spaces from the Start to 7, the student can place an object after each step along the way. The student will see that 7 objects are associated with the 7 steps from the Start to 7.

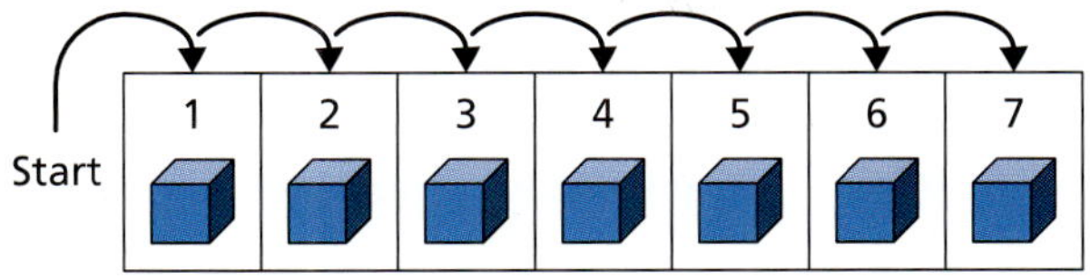

Using Ordinal Numbers to Count

Although we usually count using cardinal numbers: *one, two, three, …,* sometimes we count using ordinals: *first, second, third,….* You should discuss ordinal numbers because they are a part of students' everyday lives. There are many opportunities during the course of a day for a teacher to use ordinal language, for example, "Would the third student in each row come to the front? Who came first in the race? Who is sixth in line?"

It is important to help students see the relationship between cardinal and ordinal numbers. For example, if there are 6 people in line and we start with the first person and say "one," then we say "sixth" for person number 6.

Students should learn that when we use a word like "sixth," it suggests that there are 5 items ahead of it. You need to consider point of view when you deal with ordinals. In the picture below, the square is sixth if you look from the left, but second if you look from the right. This is very different from cardinal numbers, where position is irrelevant.

The square is sixth or second, depending on your point of view.

Counting On

When students have become confident in their counting, instead of starting to count at 1, which is often called *counting all*, they can *count on* from the number that describes a subset of known size within that set of objects. For example, students might recognize the group of 4 counters as 4 without counting (see Subitizing on **page 20**) and then count on, saying, "5, 6, 7, 8, 9, 10, 11."

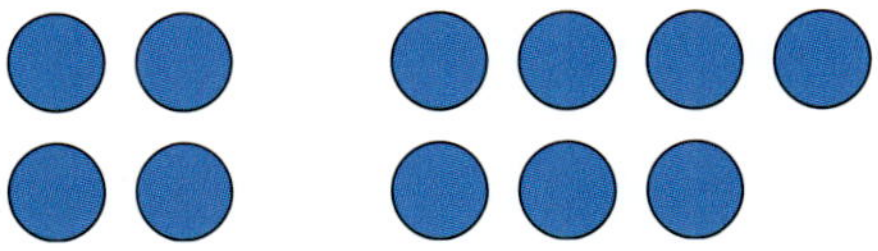

"I see 4 there, so there are 5, 6, 7, 8, 9, 10, 11 altogether."

When they are able to count on, students will learn addition combinations more easily. For example, to add 4 + 7, we want students to realize that they could start at 4 and then count on another 7.

There are a number of ways to help students practise counting on. One way is to have students stand on a particular number on a walk-on number line and then ask them to move forward several spaces, saying the numbers they walk on as they move. For example, if a student starts on 7 and has to move forward 5 spaces, she says the numbers "8, 9, 10, 11, 12." It is important

Teaching Idea 2.2

Hide 3 items in your hand or under a piece of paper. Tell students that 3 items are hidden. Put out 4 more items where students can see them. Give students additional counters to model the 3 hidden items if they need to.

To focus on BIEN 2, ask: *How many items are there altogether?* [7] Observe how the students count and then say: *Show me how you counted.* [e.g., "4, 5, 6, 7" or "1, 2, 3, 4, 5, 6, 7"]

for the student to start counting at 8 and not at 7 (a common error). If she starts at 7 and then counts on 5 numbers, she will get the wrong result. It is also helpful for students to use their fingers to keep track of how many numbers they have counted, since the 5 is not readily apparent otherwise.

Teachers can incorporate a lot of counting on into "sponge activities" (activities that you squeeze in when there is a minute or two to spare). For example, while students are packing up for the day, start saying "7, 8, 9, ..." and have students join in the chant. Many teachers emphasize the ability to say a number that is 1 or 2 more, or 1 or 2 less, than a given number. For example, they might call out the number 6 and ask for the number that is 2 more; students are expected to think, "7, 8." Some students might whisper 6, but they have to be careful not to count it or they might end up at 9 instead of 8.

Counting Back

Teachers often overlook counting back, which is an important skill to prepare students for later work in subtraction.

Teachers often overlook counting back, which is an important skill to prepare students for later work in subtraction. Some children's counting books and songs, like *Pirates at Sea* (Hewitt, 2000), or *Five Little Monkeys Jumping on the Bed* (Christelow, 1998), provide opportunities to count back. You can easily create such opportunities yourself. For example, just before starting a new activity, you can begin counting, "10, 9, 8, ..." and have students join in. It is important to start at different points, for example, sometimes 10, 9, 8, ..., but other times 15, 14, 13, ..., or 6, 5, 4, ...

Teaching Idea 2.3

Provide a number line. It could be an electronic number line like this one.

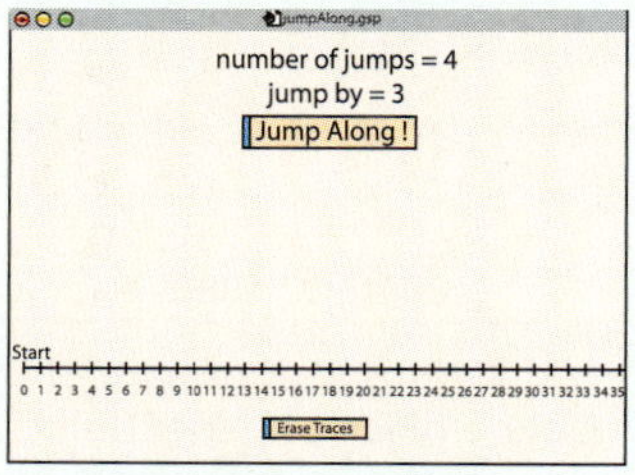

Tell students that you want them to start at the start position and get to the number 20 by making jumps of equal size. Ask them to experiment to see what jumps are possible.

To focus on BIEN 2, ask: *What numbers did you say as you made your jumps?* [e.g., 2, 4, 6, ..., 20] *I noticed that you jumped by 2s. What if you started jumping at 14 instead of at the Start; would you still get to 20?* [yes] *How do you know?* [e.g., I just tried it and it worked: 16, 18, 20.]

Skip Counting

Skip counting, a precursor to work in multiplication, introduces students to the notion of unitizing in counting. This time, groups are counted as ones. For example, when skip counting by 10s, we count two groups of 10 by saying just two numbers: 10, 20, ...

In the early years we emphasize skip counting by 2s, 5s, or 10s, and later students skip count by 100s, 50s, and 25s. The focus of skip counting in these early years is to help students see the patterns in our place value system, to prepare students for work with money, and to prepare them for work with multiplication and division.

Students learn the skip counting patterns by highlighting the numbers they hear when they skip count on a number line. Highlighting should be visual, as shown below, but it can also be physical, for example, by clapping when reaching the number that is said aloud. In either case, it is important that students observe the unitizing visually or kinesthetically.

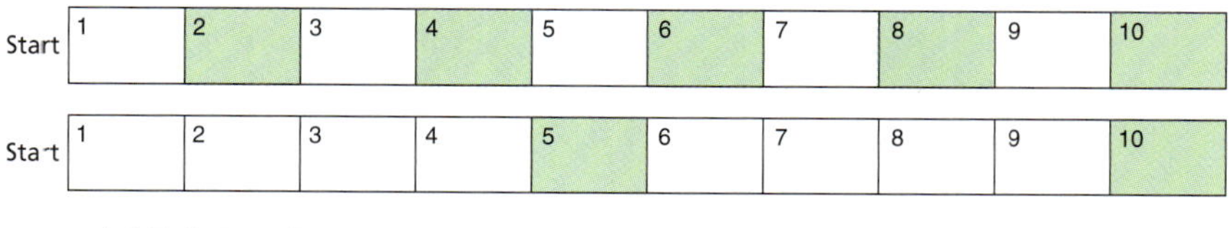

Highlighting the numbers you say when you skip count by 2s and by 5s

Over time, students should have experiences in skip counting forward and backwards. They should not always start with a typical number. For example, we often skip count "2, 4, 6, 8, ...," but not as often "1, 3, 5, 7, ...," which is also helpful to students. Beginning at higher numbers such as 7 and going both ways: 7, 5, 3, 1 and 7, 9, 11, 13, ... is also valuable.

A Sense of Number

Comparing and Relating Numbers

Many everyday situations provide opportunities for students to compare numbers. For example, they can determine whether there are enough napkins for all of the students at a table, or whether there are 2 too many or 1 too few. You can quickly create many simple comparing opportunities, for example, by asking students to clap 2 fewer times or 3 more times than you do.

Students employ these two primary strategies for comparing numbers:

- one-to-one correspondence, and
- comparing positions in the counting sequence.

Comparison Language

Whenever possible, it is preferable to say "4 is greater than 3" rather than "4 is more than 3" to teach that mathematical language to students. Eventually, they will learn to use the greater-than symbol: $4 > 3$.

Many students have difficulty recognizing that if 4 is greater than 3, then 3 is automatically less than 4; they tend to focus on the "greater than" aspect of the relationship. Make sure they consider both sides of the relationship.

Using One-to-One Correspondence to Compare

One-to-one correspondence is the basic test for comparison. Students model both numbers, matching up elements of the two sets. The set that has items left over is the larger set, and the number that represents that set is the greater number.

4 is greater than 3 since each boy matches 1 of the girls and there is 1 extra girl.

An everyday example of one-to-one correspondence revolves around interpreting simple graphs. For example, you can ask students to place a cube on a stack that represents the activity they prefer to do, as shown on the next page. They can easily see which activity more students chose.

Teaching Idea 2.4

Place students in groups. Have the students in each group write their first names on a grid, one letter per square, in a list.

To focus on BIEN 4, ask: *Which person in your group has the greatest number of letters in their name? How can you tell without counting?* [e.g., My first name has the most letters because it sticks out farthest.] *Now compare the lengths of your last names but do not use a grid. Who has the name with the greatest number of letters? How do you know?* [e.g., Jason's name has the most letters. We counted the letters in each name. Jason's last name has 7 letters and mine has 5. Since I say 7 after 5 when I count, 7 letters is longer.]

Teaching Idea 2.5

Provide students with 20 counters. Ask them to use the counters to show why 9 is greater than 7.

To focus on BIEN 4, ask: *Why might it help to make a row of 9 and a row of 7 and line them up?* [to match the counters to show that there are more in the 9 row] Ask students to explain how they know that 15 is greater than 14. Observe whether they use the counting sequence to help them compare the numbers [When I count, I say 15 after I say 14, so 15 is greater.], since there are not enough counters to model both at the same time.

My Favourite Activity

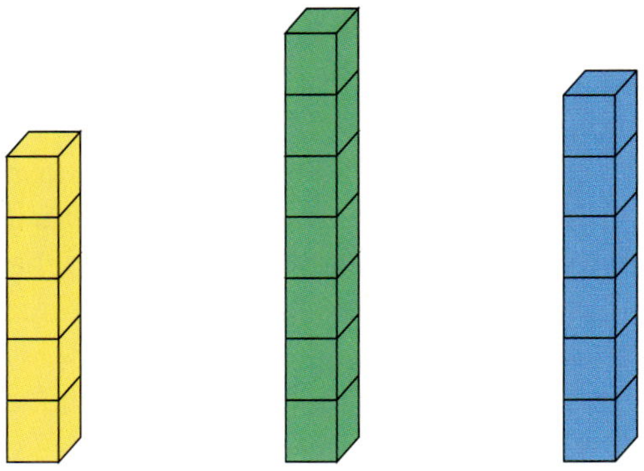

You use one-to-one correspondence to interpret simple concrete graphs like this.

Using Position in the Counting Sequence to Compare

Later, students count the items in both sets separately and then compare the position of those two numbers in the number sequence. For example, if one set has 3 items and one has 4 items, they know that since the number 3 comes before 4 in the counting sequence, 3 is less than 4, and the set of 3 contains fewer items than the set of 4.

Comparing Ordinal Numbers

Comparison is usually reserved for cardinal numbers, not ordinal numbers. For example, we say 4 is greater than 3, but we do not say fourth is greater than third. However, because ordinal numbers are meaningful to students in everyday situations, they might compare them intuitively. Ordinal numbers are interesting because the way they compare is based on context. Sometimes, being fourth is better than being third and sometimes worse. For example, in a game where the lowest score wins, if the scores are placed in order from least to greatest, the fourth score would be better or greater than the third score. Other times, the reverse is true.

Teaching Idea 2.6

To focus on BIEN 4, ask students to say which is greater, 4 or 10, and how they know. Then ask: *If 10 is greater than 4, is something that is tenth greater than something that is fourth?* [Not always; e.g., If I list the ages of 10 people from youngest to oldest, the tenth number is the greatest age, so it is greater than the fourth number. If everyone in the class reads lots of books and we list the number of books from greatest to least (since the student with the greatest number of books read came in first) the tenth number is actually less than the fourth one.]

Numbers to 10

Subitizing

Subitizing is the ability to tell how many instantly, without counting. It is immediately recognizing an arrangement or configuration of a certain number of items. There are actually two kinds of subitizing.

- Perceptual subitizing involves immediate recognition without using mathematical processes; for example, two year olds can distinguish between sets of 2 and sets of 3 without counting or having language for those quantities.
- Conceptual subitizing (Clements, 1999) is, for example, seeing two rows of four and immediately recognizing it as 8 without counting; seeing units of units.

Although rectangular arrangements are often easier for students, most people immediately recognize, or subitize, the arrangements for 1 to 5 shown at the top of the next page, as these configurations of dots are commonly used on dice, cards, and other game materials. Many students are also instantly familiar with finger arrangements for numbers like 2 to 5 as well.

Individual students' experiences will determine which numbers they can recognize immediately.

Familiar dot configurations for the numbers 1 to 5

Many teachers play flash games using paper plates with circle stickers or dots created by bingo dabbers. They show a dot arrangement very briefly and ask students if they recognize it. Although some arrangements are more commonly used than others, some teachers use a variety of representations often enough that students begin to recognize them right away. Teachers might sometimes ask students to verbalize how they recognized the number they did, particularly numbers greater than five, although that is not necessary every time.

Different configurations for the number 9

5 and 10 as Benchmarks

Many students think of one number in relationship to another number. For example, 6 is the number that is 1 more than 5, or 9 is the number that is 1 less than 10. Encourage students to relate numbers to benchmark numbers as a tool for comparing numbers (as shown below). Five and 10 often provide the strongest benchmarks for young students. They can use these benchmarks to estimate quantities (e.g., it looks like it's closer to 5 than 10) or to count (e.g., it's 2 more than 5, so it's 7).

One way to encourage students to refer to benchmarks is to use a visual model. Aside from the using the fingers on one or both hands, three other models are particularly useful: the bead number line or Rekenrek, the 5-frame, and the 10-frame.

Teaching Idea 2.7

Rather than using standard dot arrangements, consider using designs like the design below to emphasize that numbers are made up of parts. For example, this card shows that 9 = 4 + 4 + 1 or 4 + 1 + 4 or 1 + 4 + 4.

To focus on BIEN 1 and 3, ask: *How many lines are there altogether?* [9] *How many lines make up each square?* [4] *How does this help you see that 9 is made up of parts?* [There are 2 squares that each have 4 lines and there is 1 line between them. That's 4 and 1 and 4, which makes 9.]

Teaching Idea 2.8

Tell students you are thinking of a number that is close to 5. Ask them to guess the number.

To focus on BIEN 5, ask: *Do you think my number could be 20? Why not?* [No; e.g., close to 5 should be 3 or 4, or 6 or 7, but not as high as 20.]

Teaching Idea 2.9

Pull 12 beads to the left on a bead number line and ask students to tell you how many beads you have pulled over.

To focus on BIEN 3 and 5, ask: *How did knowing that a group of red beads and a group of blue beads is 10 beads help you count the beads quickly?* [I started at 10 and then counted 11, 12.] *How do the beads show that 12 = 5 + 5 + 2?* [There are 5 red, then 5 blue, and then 2 more red.]

The Bead Number Line Model Many teachers use beads to create a number line on a string or use commercial Rekenreks.

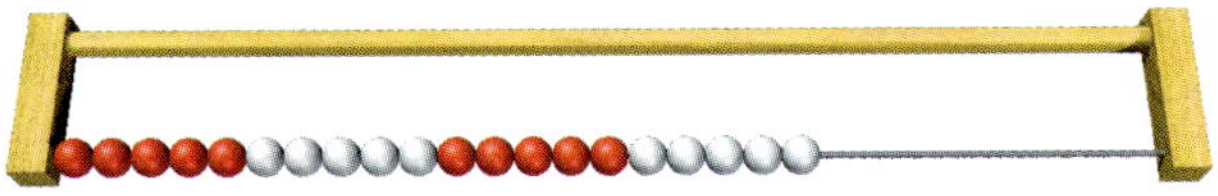

Rekenrek

To create a bead number line, you can alternate colours every 5 beads. Initially, you can mark the numbers with stickers on the beads, but later you can remove the stickers. This helps students focus on comparing to a benchmark. For example, 7 is the number that is 2 past 5. A bead number line can have 10, 15, 20, or 25 beads, and sometimes as many as 100 beads.

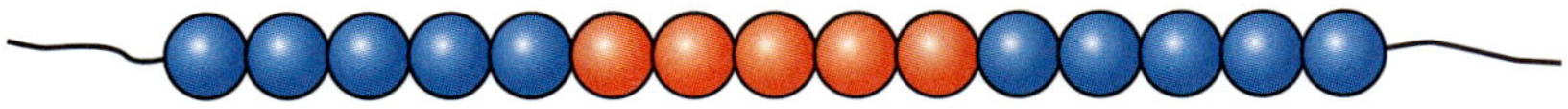

A bead number line like this helps students relate numbers to the benchmarks 5, 10, and 15.

Another variation of a bead number line is a double string, each with ten beads, five of one colour and five of the other.

The 5-Frame Model The 5-frame is a rectangle of 5 squares. Each square is large enough to hold a counter. Initially, students can place counters wherever they wish on the frame and tell how many. For example, a teacher might have a child roll a die and put that many counters on the frame. They soon see that the 5-frame can only hold numbers to five, which could lead later to the introduction of the 10-frame.

When working with 5-frame representations for 1 to 5, students learn to subitize each representation and, at the same time, think of how each number relates to 5.

Eventually, we lead students to fill the frame from the left. We help them observe how this not only makes it easier to quickly recognize how many are there because the configurations become familiar, but it also helps since, for example, filling the frames that way makes it very easy to show 4 as 1 less than 5. Students learn to subitize these representations for 1 to 5, and at the same time to think of the numbers 3 and 4 in relation to 5.

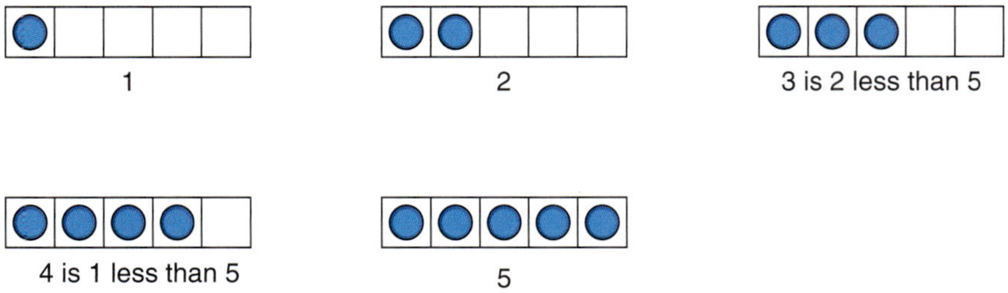

5-frame representations for the numbers 1 to 5

The 10-Frame Model The 10-frame is a set of two 5-frames in two rows. To fill the frame, you begin at the left, fill the top row, and then start at the left again to fill the bottom row. This model helps students focus on relationships to 5 and 10, for example, 6 is 1 more than 5, or 4 fewer than 10.

Some teachers use vertical 5-frames and 10-frames. In this case, the convention is to fill the first column and then the second column from the top down.

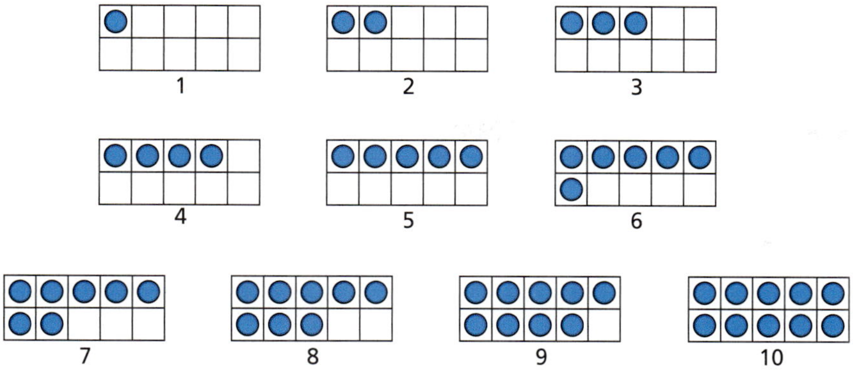

10-frame representations for the numbers 1 to 10

Once students are familiar with 10-frame arrangements, you can play a variety of games to practise the arrangements for various numbers. For example,

- Make a deck of 10-frame pictures that contains two copies of each number from 0 to 10.
- Students play in groups and take turns dealing out the cards.
- Each player starts with four cards and the remaining cards are placed face down in a stack.
- On his or her turn, each player takes the top card from the stack.
- If a player has two cards that add to 10, he or she discards them by placing them at the bottom of the stack.
- The first player to get rid of all his or her cards wins.

The Number Zero

At some point, students need to meet the number 0. There are several situations where its introduction would make sense.

One possibility is when students are learning to count back. As students are counting back and say 1, a teacher could ask: *Then what could you say?* Some students might know about 0; otherwise the teacher could introduce it. Another possibility is to give the start position on a walk-on number line a number name, specifically 0.

A third possibility is to introduce 0 as an empty 5-frame or 10-frame.

The Teen Numbers

Many students struggle with the teen numbers, 13 to 19. One of the reasons offered is that in English we say the numbers in a different order than we write them. For example, we say "sixteen" with the "six" part first and the "teen" part last, but we write it with the 1 first and the 6 last—16. Help students focus on how these numbers relate to 10. For example, 16 is sixteen because it is 6 more than 10.

Students need a great deal of experience representing and using teen numbers to become comfortable using them. It is helpful to use a double 10-frame for the teen numbers. For example, 15 as shown below is clearly 5 more than 10.

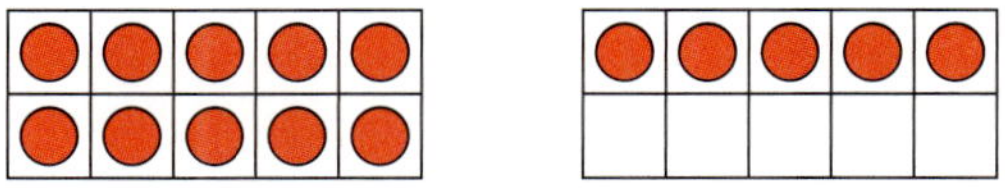

A double 10-frame for modelling teen numbers

Teaching Idea 2.10

Tell students that the number of rooms in a house is 2 more than 5.

To focus on BIEN 5, ask: *Is the number of rooms more than 10 or fewer than 10? How do you know?* [Fewer; e.g., 10 is 5 more than 5 and the number of rooms is only 2 more than 5.] *How many fewer?* [3] *How can you show this on a 10-frame to help you figure this out?* [On a 10-frame, 2 more than 5 is a whole row and 2 more of the next row, so I need 3 more to make 10, or fill the frame.]

Teaching Idea 2.11

Invite children to show 5 in different ways, using materials such as 10-frames, square tiles, toothpicks, and buttons.

To help students focus on BIEN 3, ask: *Which representation of 5 was the best way to show that 5 is an odd number?* [e.g., 2 rows of 2 squares and then 1 more square] *Which was the best way to show that 5 is half of 10?* [e.g., one row of a 10-frame]

Teaching Idea 2.12

Read the book *10 for Dinner* by J.E. Bogart (1989) to students. The pages provide an interesting setting for students to think about combinations for 10. Ask students to act out their own scenarios that would fit the story.

To focus on BIEN 3, ask: *How do your stories show that there are different ways to represent the number 10?* [The total was always 10, but there were different ways to show it in parts.]

The Importance of Multiple Representations

One of the important concepts young students need to learn about number is that there are many ways to represent an amount. For example, the number of arms you have can be represented by the symbol, or numeral 2, but also by thinking of it as 1 more than 1, or 1 less than 3 or, later, as $7 - 5$.

Multiple representation can and should be approached in a variety of ways. For example, to model the number 7, students might draw or show any of these models:

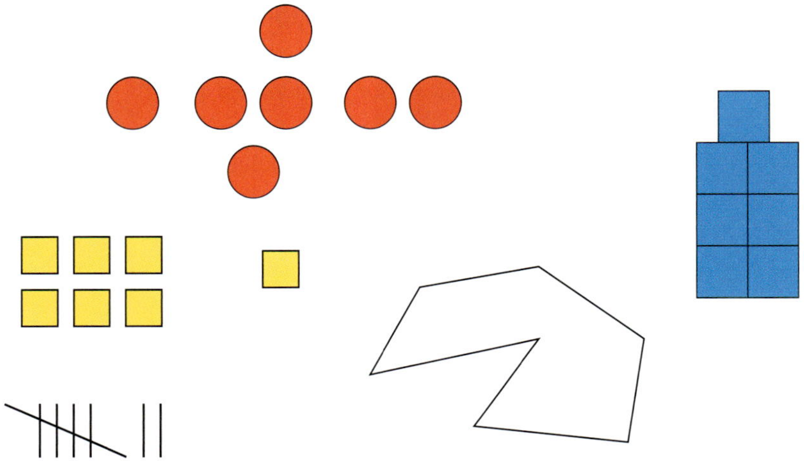

Multiple representations for 7

You might post a flip chart with three columns: one for a numeral, one for pictorial representations of the numeral, and one for the number word. Students can refer to the chart regularly.

The 6 and 1 model using yellow squares shown above uses what is called a part-part-whole relationship; students view 7, the whole, as being made of two parts, the 6 and the 1. These part-part-whole relationships are the basis for later addition work. For example, to show the different combinations that add to 7, a student rearranges 7 by separating it into parts ($7 = 1 + 6$; $7 = 2 + 5$; $7 = 3 + 4$; …). To lay a foundation for subtraction work, you can show students part of a whole and ask them to figure out the other part to make the whole. For example, show a 10-frame with 7 counters and 3 empty spaces.

It is particularly important to introduce the equals sign as a symbol that tells that two sets of symbols are representations for the same amount, an important first connection to algebraic thinking. For example, when you write $3 + 1 = 4$, you are saying that 3 and 1 is another way to represent 4. Similarly, when you write $3 + 2 = 4 + 1$, you are saying that 3 and 2 is another way to represent 4 and 1. This is often modelled with a pan balance. If, for example, you place 3 green cubes and 2 blue cubes on one pan, they will balance 4 red cubes and 1 yellow cube on the other pan.

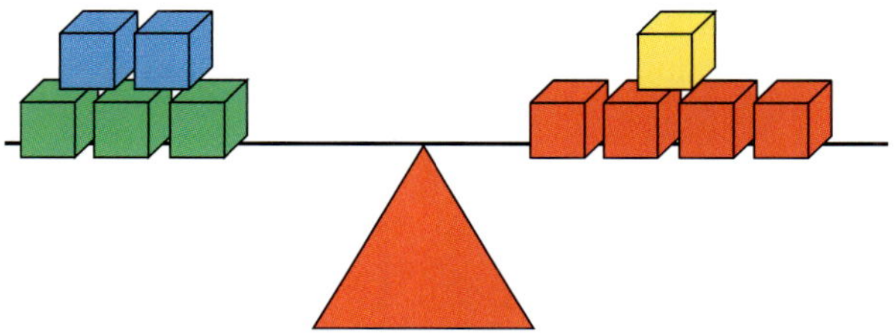

This pan balance model shows multiple representations for 5: $2 + 3$ and $1 + 4$.

Reading and Writing Numerals

Numerals are a universal way of representing numbers symbolically. The number is the actual amount that is being represented; the numeral is just the symbolic representation of it.

Reading Numerals

Children encounter numerals in many everyday situations. For example, a child might receive a birthday party invitation with numerals that represent different things.

We use numerals to represent age, time, and location, and for identification.

It is important for students to recognize numerals and what they represent, and to learn to use them. Storybooks are a great way to learn to read numerals; students can attend to page numbers and to the numbers in the stories themselves.

Many students have little trouble becoming familiar with some numerals, but much more difficulty with others. For example, they might have difficulty distinguishing numerals that share characteristics, for example, 6 and 9 both have a closed loop and a curved line.

In our digital world, students often see block form numerals rather than the handwritten forms of numerals that are often taught. It is important that students recognize the variability in representing numbers. For example, you can show them all of these different versions of the numeral 8, pointing out how they are similar and how they are different.

It is important to expose students to a variety of numeral styles.

Teaching Idea 2.13

Show a group of 6 items, such as 6 feathers. Ask students to find as many ways as they can to show the number 6. [e.g., the numeral 6, the word six, 6 counters in a row, 6 dots in 2 rows of 3]

To focus on BIEN 3, ask: *Which of the ways you showed is the quickest way for someone else to know that it shows 6? Why do you think it is quickest?* [e.g., the numeral 6, since you don't have to count or add to figure out how many there are]

You can also create number puzzle cards that children can put together to form numerals, as shown here.

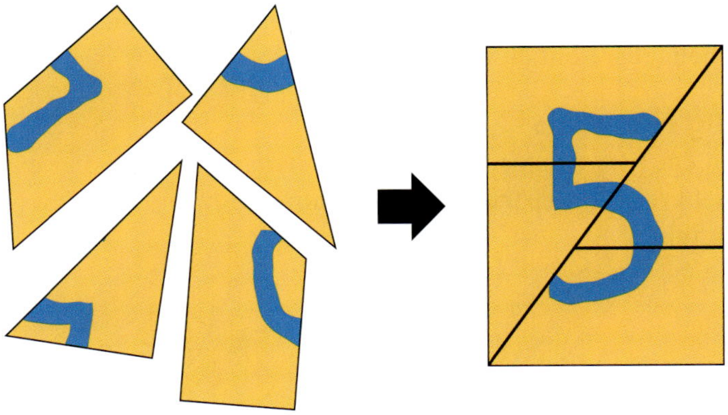

A puzzle for the numeral 5

To help students get used to numerals that look digital, provide a calculator, cards that show handwritten numerals, and cards that show stylized numerals. Ask students to press a number button on the calculator and then find the numeral cards that match what they see.

Writing Numerals

Young students will need practice in learning to write numerals as well. They need to have a starting point for writing the numeral as well as a plan for continuing it. There is no one right way to write numerals. For example, some children find it easier to write the numeral for eight as two circles (almost like a snowman), whereas others are comfortable with a single continuous motion.

There are many strategies that are helpful to practise numeral writing including

- using sandpaper numerals that students can trace with their fingers and feel,
- having students write numerals in materials like sand so that they can both feel what they're doing and also so that they can easily fix mistakes,
- having students trace numerals in the air, following teacher instruction,
- making modelling clay or edible cookies in the shape of numerals, and
- having students use follow-the-dot patterns, many of which are freely available on the Internet.

When possible, numerals should be written for some purpose, rather than just for the sake of writing them. For example, writing the score in a game or the number of the student's bus.

Greater Whole Numbers

Students need strategies to represent and make sense of greater numbers. Although it is possible to count, say, 87 items individually, it is not practical. Counting is easier and more accurate when we group items.

This becomes very clear if you put, for instance, 35 items on an overhead projector in a disorganized fashion, allow students to look for a few seconds, and then turn off the projector. Very few students can tell how many items there are. In contrast, if you show the 35 items in 3 groups of 10 and 5 more (as shown below), students find it easier to identify the amount, using conceptual subitizing.

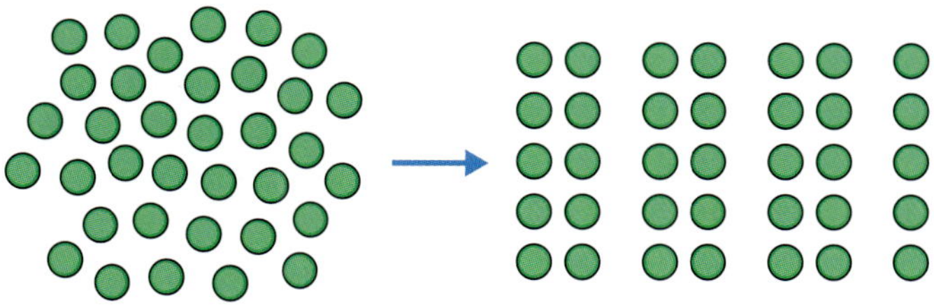

When students start working with greater whole numbers, grouping becomes a necessary strategy.

BIG IDEAS FOR GREATER WHOLE NUMBERS

1. A number tells how many are in a group. To count the number in a group, we often create subgroups and count the number of subgroups. (BIGWN 1)
2. The place value system we use is built on patterns to make our work with numbers more efficient. (BIGWN 2)
3. You can represent a number in a variety of ways. Each representation of a number can focus on a different aspect of the number. (BIGWN 3)
4. To compare and order numbers, we can compare them to more familiar benchmark numbers. (BIGWN 4)
5. Students gain a sense of the size of numbers by comparing them to meaningful benchmark numbers. (BIGWN 5)

Note that some of the big ideas listed here are identical or similar to the ones for Early Number. This makes sense since these are big ideas that underpin work with all whole numbers.

Each teaching idea in this section of the chapter will indicate which Big Idea(s) for Greater Whole Numbers (BIGWN) can be emphasized.

A Base Ten Place Value System

Much of the work in Grades 1 to 3 is designed to move students from counting individual items toward strategies that involve using groups of ten to figure out how many.

Our base ten system is arbitrary, although scholars hypothesize that we group in tens because we have 10 fingers. Humans could also have developed a base eight system since we have 8 fingers, excluding thumbs. The symbols we use could look different than the 0 to 9 digits in our current system. In spite of these choices, many of our processes for arithmetic would still be the same. If students are curious, you should help them understand that the decision to use tens is just that, an arbitrary decision. It is easier to

Teaching Idea | **2.14**

Provide packages of exactly 52 counters to groups of students. Ask them to count the counters.

To focus on BIGWN 1, ask: *What did you do to make it easier to count?* [e.g., We made groups of 10, and then we counted 10, 20, 30, 40, 50, 51, 52.] Allow time for students to share their different approaches.

communicate about mathematics if we all use the same system. We use a standard system for numbers just like we use standard units of measurement.

It is important for students to understand why we group at all, whether in tens or in eights. Activities like **Teaching Idea 2.14** and the activity shown at the top of **page 27** with the 35 counters help students understand the value of grouping. When you make groups of the same size, it is easier to tell how many.

Numeration Principles

The most basic principles that students need to learn in order to work with our number system relate to the base ten trading rules and the concept of place value:

- Base ten is a trading system in which you trade groups of 10 ones for 1 ten, 10 tens for 1 hundred, and so on.
- Place value is a system in which the position that a symbol (a digit 0 to 9 in the case of our numeration system) occupies in a numeral indicates the value of that symbol. For example, the digit 2 in the numeral 23 means 20, but the digit 2 in the numeral 42 means 2.

A number of researchers have expressed caution about moving young children into place value too quickly.

A number of researchers, including Kamii (1986), have expressed caution about moving young children into place value too quickly. Kamii points out that a number of students read the number 16 and really have no sense of what the 1 really means; they often treat it as 1 rather than as 10. She suggests teaching numbers more holistically for a longer time, for example, thinking of 23 as twenty-three, not as 2 tens and 3 ones, until students are ready for the transition, which may be later than some teachers think.

Once a teacher chooses to introduce place value, it is critical that students use concrete materials with 2-digit numerals before working more symbolically. It is also advisable to begin work with 3-digit numerals with concrete materials before working symbolically. Students should not be rushed to greater numbers until they are ready to make sense of them.

Even once students understand the place value system and are comfortable with numbers in the 100s or even 1000s, it may take much more time for them to gain a full understanding of the structure of the system going farther to the left (to the ten thousands and beyond) or to the right (to decimal values).

The base ten place value system is defined by conventions; these conventions are not implicit in what number is all about. As discussed earlier, these conventions are based on arbitrary decisions about group size and symbols. Because they are conventions, they must be taught explicitly. Some of the principles underlying our base ten system that students need to learn are listed below.

NUMERATION PRINCIPLES

1. We group in tens for convenience so that we need only 10 digits (0 to 9) to represent all numbers.
2. Patterns are inherent in our numeration system because each place value is 10 times the value of the place to the right.
3. A number has many different forms. For example, 123 is 1 hundred, 2 tens, 3 ones, and it is also 12 tens, 3 ones.
4. A place value system requires a symbol for a place holder. For example, the 0 in 304 is a place holder; it pushes the digit 3 over to show that it represents 300 instead of 30.
5. You can compare numbers when they are written in standard, or symbolic, form.

PRINCIPLE 1 We group in tens for convenience so that we need only 10 digits (0 to 9) to represent all numbers.

Students quickly realize how convenient the base ten system is if they are asked to memorize separate symbols for, say, the first 25 different numbers instead of just recognizing 10 digits. But it takes time for students to internalize the structure of the system.

Many young students have difficulty thinking about the one item, or symbol that represents a 10 as being worth more than the one item, or symbol representing 1. For that reason, it is important to use models that are proportional, that is, where the 10 is 10 times as big as the 1, the 100 is 10 times as big as the 10, and so on. Some of these models, representing 23, are shown below. In the case of the Popsicle sticks, the symbol 2 in 23 is represented by 2 bundles of 10 sticks and the symbol 3 is represented by 3 loose sticks. Note that the proportional nature of these concrete models will also help with number sense because 23 is literally bigger than 12 and smaller than 57.

PROPORTIONAL MODELS FOR REPRESENTING 23 CONCRETELY

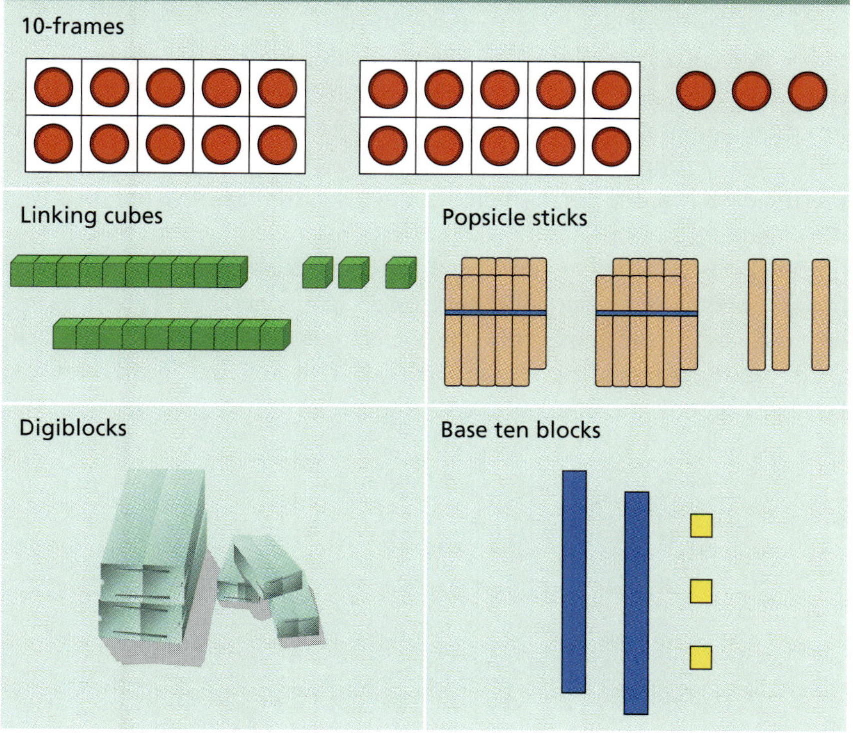

Other models that can be used to make the trading meaningful can be based on context. For example, if boxes hold exactly 10 cans, it makes sense that it takes 4 full boxes and 3 loose cans to show 43 cans. No matter which models are used, it is important for students to have practice trading up and trading down, for example, trading 28 ones for 2 tens and 8 ones or the reverse.

Although all of the models presented above can be extended in one way or another to numbers greater than 100 (e.g., bundles of 10 bundles of popsicle sticks, sheets with ten 10-frames, and so on), for practical reasons usually only the base ten block model is extended. The base ten blocks are convenient, but one difference between the base ten blocks and other models is that, for many base ten blocks, students cannot literally take apart a ten block (a rod) to make 10 one blocks (units) or a hundred block (a flat) to make 10 ten blocks (rods). Therefore, if the base ten block model is introduced too early, students often do not realize that the value of, say, 2 rods

Teaching Idea 2.15

An engaging context for young students to think about grouping items in tens is a roller coaster with cars that hold 10 people each. You can ask students to fill empty roller coaster cars to show various numbers of people. Or, students can describe how many people there are altogether in cars that you model.

To focus on BIGWN 2, ask: *If there were 8 riders, we would use one symbol, 8, to show the number of riders. Suppose there were 28 riders. Why do we use two symbols, a 2 and an 8, to show that number of riders?* [The 2 shows there are 20 because it means 2 tens and the 8 shows there are 8 more because it means 8 ones.]

Teaching Idea 2.16

An engaging activity for young students to focus on the idea of trading is this simple game. Give each student a mat like the one shown below. Students take turns rolling a die and taking that many yellow counters, which they place in the yellow column of the mat. Once they have 3 of one colour, they trade for 1 counter of the next colour (e.g., 3 yellows for 1 blue, 3 blues for 1 green, 3 greens for 1 red). Ask students what colours they would have if they had 2 blues and 2 yellows and rolled a 4 on the die. [1 green and 1 blue]

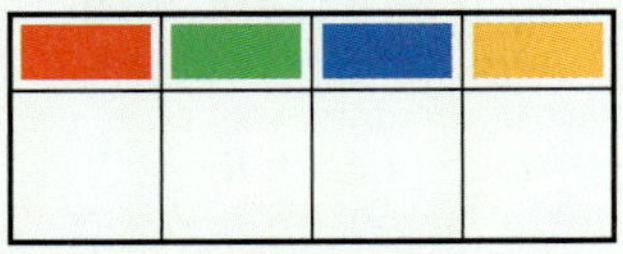

To focus on BIGWN 1, ask: *Why would you never have more than 2 counters in a column after you did your trading?* [If I had 3, I would trade for the next colour.]

To focus on BIGWN 2, ask: *Suppose you couldn't trade until you had 10 of one colour. You have 2 blue and 1 yellow. You roll two dice and get a total of 10. What colours would you have now?* [3 blue and 1 yellow] *Why is it easy to figure out?* [10 more yellow means 1 more blue.]

Teaching Idea 2.17

Tell students that you are thinking of a number that you would say as "something hundred forty-nine," but do not say how many hundreds.

To focus on BIGWN 2, ask: *What number comes next?* ["something hundred fifty"] *How can you answer without knowing what the entire number is?* [It doesn't matter how many hundreds there are in a number—50 always comes after 49.]

and 3 unit cubes is the same as the value of 23 unit cubes. That is why it is important in the early grades to have students build the materials themselves. Popsicle sticks, linking cubes, or 10-frames as models all allow for this.

Later, students can use non-proportional materials, such as different coloured counters on a place value mat or coins like loonies, dimes, and pennies (sometimes called a value model). The place value model below shows 314, but it is not obvious that the value of each red counter is 10 times as much as a yellow counter and 100 times as much as a blue counter.

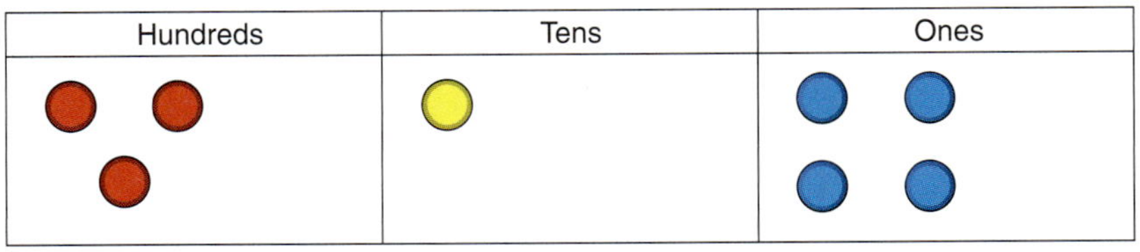

A non-proportional place value model for 314

PRINCIPLE 2 Patterns are inherent in our numeration system because each place value is 10 times the value of the place to the right.

Many students easily take to the rhythm of the number system that is the result of the base ten relationships (once they get past the teen numbers), for example, twenty-one, twenty-two, ..., twenty-nine, thirty, thirty-one, thirty-two, ..., thirty-nine, forty. Or one, two, three, ..., one hundred one, one hundred two, one hundred three, ..., two hundred one, two hundred two, two hundred three, ... The 100 chart is a useful tool to help students recognize patterns for numbers up to 100. Some teachers prefer charts that go from 0 to 99; in these charts, all the numbers in any one row share the same tens digit and all the numbers in any one column share the same ones digit. Other teachers prefer charts that go from 1 to 100, mostly because the number 1 feels like a more natural starting point.

0	1	2	3	4	5	6	7	8	9
10	11	12	13	14	15	16	17	18	19
20	21	22	23	24	25	26	27	28	29
30	31	32	33	34	35	36	37	38	39
40	41	42	43	44	45	46	47	48	49
50	51	52	53	54	55	56	57	58	59
60	61	62	63	64	65	66	67	68	69
70	71	72	73	74	75	76	77	78	79
80	81	82	83	84	85	86	87	88	89
90	91	92	93	94	95	96	97	98	99

"Twenty-eight, thirty-eight, forty-eight, ..., ninety-eight"

Another important pattern in our system is the base ten relationship between the values of the digits in the numeral, which allows us to interpret the value of an unfamiliar numeral. Each place value is 10 times the value of the unit to its right and $\frac{1}{10}$ of the value of the unit to its left. The base ten block models typically used to represent whole numbers are designed to show this pattern. For example, each 100 flat is 10 times the value of each 10 rod and $\frac{1}{10}$ the value of each 1000 cube.

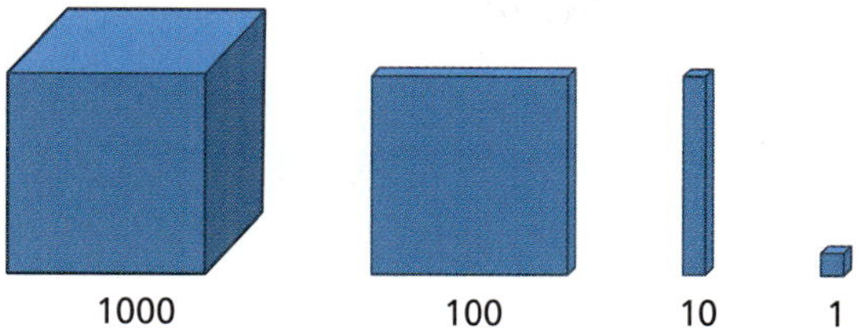

Base ten blocks concretely show the base ten relationship between the values of the digits they represent.

PRINCIPLE 3 A number has many different forms.

A place value system is built on the notion that a number has many forms. It is because 37 ones can be regrouped as 3 tens and 7 ones that we know to say thirty-seven. It is also useful to use different forms of a number to get a sense of its size. For example, to compare 72 and 60, it is helpful to write 72 as 60 + 12 rather than as 70 + 2.

The ability to rename numbers is fundamental to many of the algorithms students will learn that involve addition and subtraction. For example, regrouping 3 hundreds, 1 ten, and 2 ones as 2 hundreds, 10 tens, and 12 ones makes the traditional subtraction algorithm possible for a calculation such as 312 − 178, as it allows you to subtract 8 ones from 12 ones and 7 tens from 10 tens (see Algorithm 1 on **page 55**).

PRINCIPLE 4 A place value system requires a symbol for a place holder.

When you use digits to write a number in its symbolic form, for example, 304, the digit 0 is a place holder. If you did not have the digit 0 to hold the tens place, the number would be recorded as 34 and you would mistakenly think that the 3 represents 30 instead of 300.

If you always recorded the digits of a numeral on a place value mat or modelled it with base ten blocks, as shown below, the digit 0 would not be necessary. It is only for symbolic recordings that you need 0 as a place holder.

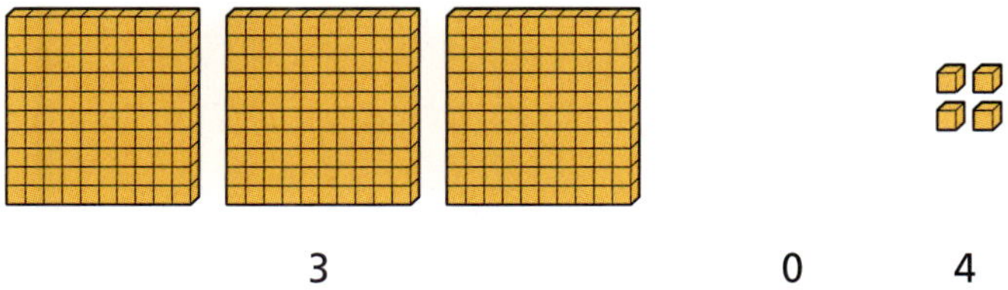

You only need the digit 0 as a place holder when you represent a number as a numeral.

PRINCIPLE 5 You can compare numbers when they are written in standard, or symbolic, form.

To learn to write numbers in standard form, students must realize that the order in which numbers appear matters. That is, even though 53 and 35 have the same digits, they mean two very different things. Some students need more time than others to reach this understanding. You can help by continuing to work on place value mats with blocks or digit cards as long as students find them helpful, and by making it explicit that the greater place values always appear to the left from the students' point of view.

Teaching Idea 2.18

A powerful way to use the 100 chart is to cover some numbers and have students use the patterns to figure them out by focusing on small parts of a 100 chart, for example,

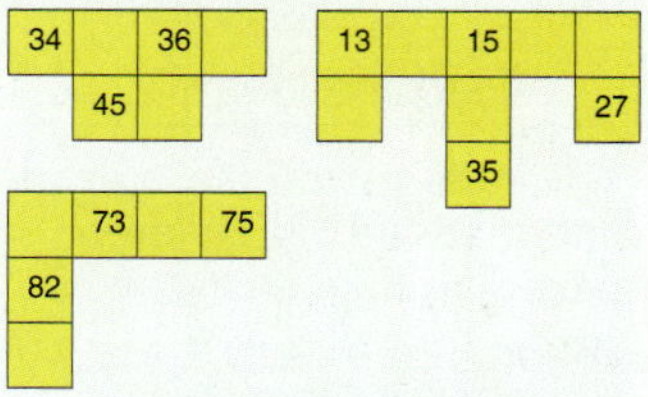

To focus on BIGWN 2, ask: *Why did you not need the whole chart to find the missing amounts?* [The patterns in the chart work for each section of the chart—the ones digit goes up by 1 as you go to the right and the tens digit goes up by 1 as you go down.]

Teaching Idea 2.19

Start writing a list of numbers: 800, 801, 802, 803, ... Ask students to think about how the pattern will continue.

To focus on BIGWN 2, ask: *How many numbers will you write altogether before the first digit is not an 8? How do you know?* [100; There are 100 numbers in each group of hundreds and this is the group of 800s.]

Teaching Idea 2.20

Ask students to use 5 base ten blocks to represent a number greater than 20, such as 41, and then write the number in the form ___ tens + ___ ones. Then ask them to represent the same number in different ways with different numbers of blocks, describing each in the form ___ tens + ___ ones.

To focus on BIGWN 3, ask: *Choose two representations of 41. Tell something that is easier to see about the number 41 in one representation than in another representation.* [e.g., 4 tens + 1 one shows that 41 is very close to 40, but 2 tens and 21 ones does not show it as well.]

When whole numbers are represented in their standard, or symbolic, form, students can use the number of digits to get a sense of their size in order to compare them. For example, any 3-digit whole number is 100 or greater, but less than 1000. Similarly, any 2-digit whole number is less than 100. Therefore, any 3-digit whole number is automatically greater than any 2-digit whole number. They can also compare numbers with the same number of digits by attending to the digits that have the same place value.

Describing and Comparing Numbers

It is important to provide students with many opportunities both to describe and to compare numbers.

Describing Numbers

To describe a number, students might use different representations of the number, different models for the number, situations in which the number might be used, and perhaps another number that is close by. For example, a student might describe 123 as 100 + 20 + 3, as a number that can be modelled with 1 flat, 2 rods, and 3 ones, as the number of students in his or her school, and as a number that is close to 125.

Comparing Numbers

One-to-One Matching

To compare numbers concretely, students can sometimes use one-to-one matching, although the numbers involved can make it awkward. For example, matching is a useful strategy to compare 123 and 112 concretely, since when you match hundreds, tens, and ones, there is 1 extra ten and 1 extra one in 123. However, it is harder to compare 217 and 145 in this way, since one number has more hundreds but the other numbers has more ones and tens.

Using One-to-One Correspondence to Compare Concretely

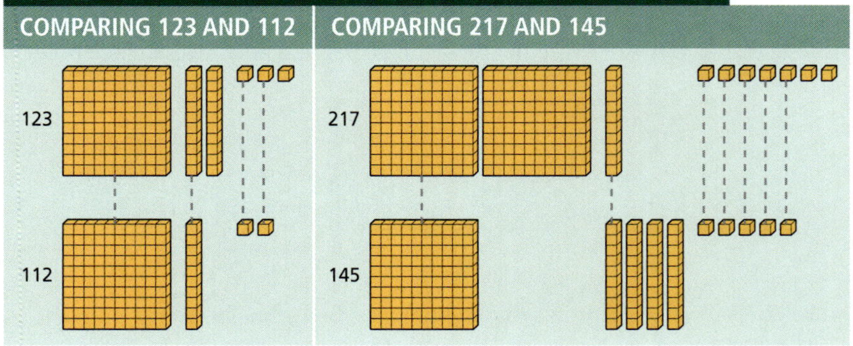

Using a Number Line

Students can compare numbers pictorially, using a number line. They compare each number to known benchmarks on the line to plot the numbers and then they use the position of the numbers to decide which is greater.

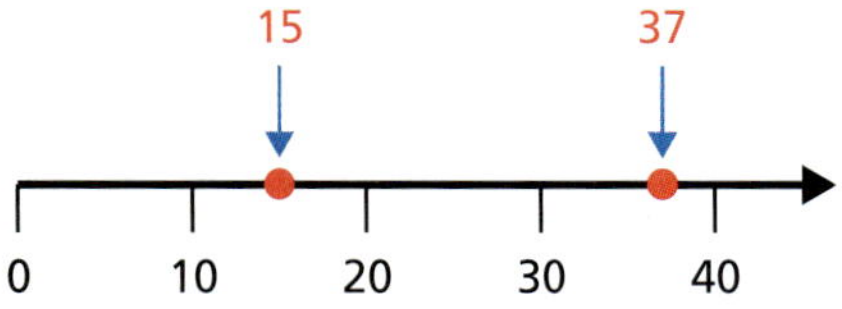

15 is halfway between 10 and 20.
37 is between 30 and 40 but closer to 40.
37 > 15 since 37 is to the right of 15.

Teaching Idea 2.21

Tell students you saw a headline that said 352 children participated in an event.

To focus on BIGWN 5, ask: *Do you think that number of children would fit in a classroom? in the gym? How do you know?* [No; e.g., when we have the other Grade 3 class in our classroom with us, it's almost full; Yes, in our school, there are 300 students in 10 classes and, when we sit in our gym, there's room for maybe 2 or 3 more classes, and 52 is about 2 classes.]

Teaching Idea 2.22

To focus on BIGWN 3, have students represent the number 175 in different ways by thinking about money. For example, you could ask: *How can you show 175 cents using quarters?* [7 quarters] *How can you show it using loonies and quarters?* [1 loonie and 3 quarters] *How can you show it using dimes and pennies?* [e.g., 17 dimes and 5 pennies] *Which way of showing 175 makes it easiest to see that 175 is an odd number? Why?* [e.g., 6 quarters, 2 dimes, and 5 pennies; I can pair up matching coins except for the last penny.]

Students eventually start to see that a 3-digit number with more hundreds than another 3-digit number or a 2-digit number with more tens than another 2-digit number is automatically greater. Rather than teaching this as a rule, you should help students understand why. One approach is to compare both numbers to a common benchmark, as shown below. If one number is below a particular value and the other number is above it, the number that is above must be greater. For example,

423 > 43 because 423 > 100 and 43 < 100.

523 > 489 because 523 > 500 and 489 < 500.

Comparing Digits Starting at the Left

Another more typical approach involves comparing digits with the same place value. Again, it is important for students to understand why the rules work. For example, the numbers 459 and 472 both have the same hundreds digit, so you compare the tens digits. This approach is reliable because the 9 ones in 459 are not enough to make another ten, much less another 20, so 472 must be greater than 459.

If two 3-digit numbers have different hundreds digits or two 2-digit numbers have different tens digits, it is easy to compare the numbers. For example, you can compare 256 and 415 easily—since 400 > 200, then 415 > 256. The 56 in 256 is not enough to make another 100, much less another 200, so 415 must be greater than 256.

We want students to realize that when two whole numbers have the same number of digits, the leftmost digit in each matters most when you compare the size of the numbers because that place has the greatest value. Another way for students to think about the role of the various digits in a number is to realize that there are digits that are more and less important. For example, in 321, the 3 is quite important; it really gives a sense of the size of the number. The 1 is also important, but for different reasons. For example, it shows that the number is odd, but it does not help at all with respect to the overall size of the number.

Counting Based on Place Value

Place value work beyond 100 depends on students' ability to count using more than one grouping in the same situation. When you count by ones, twos, fives, or even hundreds, you are counting in a consistent way. But when you count to, for example, 223 by saying 100, 200, 210, 220, 221, 222, 223, this is a complicated form of counting. Students must learn that they can start counting by 100s, and then by 10s, and then by 1s.

Another stumbling block for students in counting greater values is the transition from numbers like _99 to _00 and from _ _99 to _ _ 00, where a chain reaction of changes happens to each digit. You can support students by posing problems like this: *The population of a small community is 298. A family with several children moves into town. What is the population now?*

Estimating Numbers

Often you do not require an exact amount to represent a number; an estimate is sufficient. Students should be comfortable with a variety of situations where estimating is done, and with multiple ways to estimate numbers, depending on the context and the numbers involved. As students share their estimates, you can gain insight into their number sense.

Teaching Idea 2.23

Ask students to draw a number line from 300 to 400 and place the number 323 on the line.

To focus on BIGWN 5, ask: *How did you decide what other numbers to put on the number line? How do those numbers help show 323?* [e.g., 325, 350, 375; It shows that 323 is almost 325, so it is one fourth of the way between 300 and 400.]

Teaching Idea 2.24

Ask students to model 41 and 29 with base ten blocks and then count the number of blocks they used for each. Ask: *Can you use the number of blocks to decide which number is greater? Why not?* [No; The ten blocks have 10 times the value of the one blocks, so you can't compare how many blocks there are altogether.]

To focus on BIGWN 4, ask: *What is a better way to compare 41 and 29?* [e.g., Since 41 > 40 and 29 < 30, I can compare 40 and 30 instead.]

Teaching Idea 2.25

Some students struggle with the idea that you can estimate the same number in different ways. They also have trouble if the estimation is not meaningful, for example, when applying the rounding conventions to estimate in situations such as rounding numbers like 49 and 51 to the nearest hundred.

To focus on BIGWN 5, ask: *What numbers would you round to 100? Would you round 107 to 100?* [e.g., I would only round numbers between 95 and 105 to 100; Yes, if I was adding 107 to other numbers that were all around 200 or more.]

The three main situations where estimation is used are

- in computation; for example, to estimate 35 + 27, you could estimate 35 as 30 and 27 as 25 to estimate 30 + 25 = 55
- to get a sense of the size of numbers in order to compare them; for example, 389 is close 400 and 315 is close to 300, so 389 > 315
- for reporting numbers; for example, about 450 people came to the event

Rounding

One approach to estimating is to round a number to the nearest multiple of 10 or 100. The rounding conventions can be introduced by having students think of a number line as a highway with gas stations at the multiples to which a number could be rounded. The number that you round to is the location of the closer gas station. For example,

- when rounding to the nearest ten, the gas stations appear at multiples of 10. The nearest gas station to 133 is at 130, so 133 rounds down to 130.

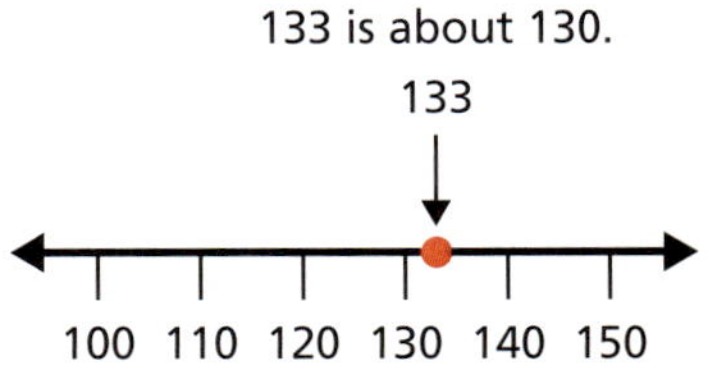

Following the rounding conventions, 133 rounds to 130 because 133 is closer to 130 than to 140.

- when rounding to the nearest 100, the gas stations would be at 0, 100, 200, 300, ... The nearest gas station would be at 100, so the convention is for 133 to be rounded down to 100.
- when rounding a number like 135 to the nearest 10, there would be two gas stations that are equally close, at 130 and at 140. Since you would have to travel the same distance to get to either gas station, the convention is to choose the one that is on your way, so 135 would round up to 140.
- when rounding a number like 150 to the nearest 100, you can apply the same logic as above. There would be two nearest gas stations, at 100 and at 200, and the convention would be to choose the one at 200, so 150 would round up to 200.

For the purposes of estimating numbers, the rounding conventions no longer apply—the numbers are "rounded" according to the numbers involved and the context or problem situation.

Note that, for the purposes of estimating numbers, the conventions no longer apply and numbers are "rounded" according to the numbers involved and the context or situation. For example, if you were estimating 135 + 147, you might round 135 down and 147 up to 130 + 150, since rounding both up would result in a very high estimate.

Benchmark Numbers

You can also estimate using other benchmark, anchor, or comfortable numbers. For example, many people are comfortable with the 25s, likely because of our monetary system. For example, a principal might estimate a class with between 23 and 28 students as having about 25 students instead of estimating it as 20 or 30.

Whole Number Operations

The four operations that are the focus of K to 8 mathematics—addition, subtraction, multiplication, and division—are highly related. These relationships form the basis of the first big idea for Whole Number Operations.

BIG IDEAS FOR WHOLE NUMBER OPERATIONS

1. There are relationships between the four operations: (BIWNO 1)
   - You can represent subtraction as the opposite of addition and vice versa.
   - You can represent multiplication as repeated addition and vice versa.
   - You can represent division as repeated subtraction and vice versa.
   - You can represent division as the opposite of multiplication and vice versa.
2. There are many situations to which an operation is applied, and there are many procedures, or algorithms, for each operation. (BIWNO 2)
3. Operation procedures should be taught meaningfully, taking into account the various meanings of the operations and the principles that apply to their use. (BIWNO 3)
4. A personal "invented" algorithm is often more meaningful and sometimes equally efficient as a conventional algorithm. (BIWNO 4)
5. There are a variety of appropriate ways to estimate sums and differences depending on the numbers involved and the context. Estimates are useful for both checking calculations and because sometimes an estimate is all that is needed. (BIWNO 5)

Each teaching idea in this section of the chapter will indicate which Big Idea(s) for Whole Number Operations (BIWNO) can be emphasized.

Relating the Operations

The diagram below shows the most critical aspects of the relationships among the four operations.

How the Four Operations Relate

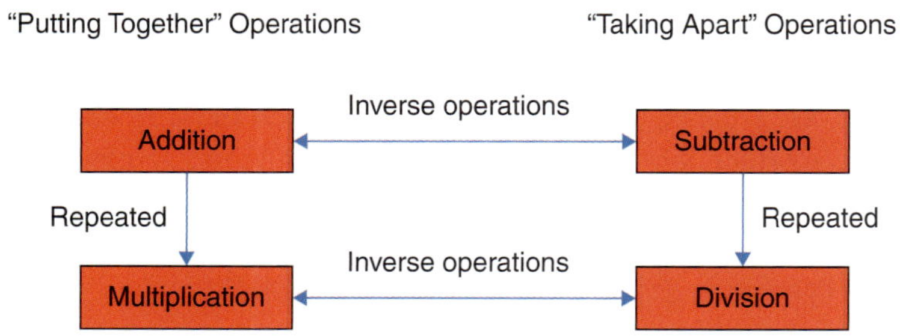

Because of these relationships, you can often solve a single problem using any of the four operations. For example,

The more students see the connections between the operations, the more successful they will be in mathematics.

PROBLEM	OPERATION	SOLUTION	ANSWER
Rafi has 12 cookies on a plate. If he eats three cookies at a time, how many times can he go back for more cookies?	Addition	I'll add 3s until I get to 12: $3 + 3 + 3 + 3 = 12$ I added 3 four times.	4 times
	Subtraction	I'll subtract 3s until there are 0 cookies left: $12 - 3 - 3 - 3 - 3 = 0$ I subtracted 3 four times.	4 times
	Multiplication	I'll figure out what to multiply 3 by to get 12: $\underline{4} \times 3 = 12$	4 times
	Division	I'll divide 12 by 3: $12 \div 3 = \underline{4}$	4 times

Addition and Subtraction

In Canadian schools, teachers generally introduce addition and subtraction formally before they introduce multiplication and division. A solid foundation in the various meanings and principles of addition and subtraction is vital for developing students who are flexible and proficient at doing the calculations that are essential to problem solving.

Early Addition and Subtraction Situations

In Kindergarten and Grade 1, teachers offer problems for students to solve that might involve addition and subtraction, but often without using any formal addition or subtraction language. In fact, the problems may seem like counting problems to the students, not addition and subtraction problems, particularly since they tend to solve these problems by counting. An example might be a teacher suggesting that there were three students who had signed up for an activity and now four more had signed up. The teacher might ask, *How many have signed up altogether? How many more students can sign up if there is only room for 12?*

Most young students will use some sort of model, such as counters, to help them solve the problem. It is only later on that formal symbolism and language are applied to the situation, for example, a teacher asking if the students had added or subtracted to determine each result. The point of including that formal language is only to simplify communication about the general principles of addition.

Teachers should attend to whether the introduction of symbolism seems to assist the students or whether it actually inhibits them.

Teachers should be aware, though, that the use of less formal language should not be viewed as problematic. It is certainly not important that Kindergarten students write equations such as $3 + 4 = 7$. Although these equations are typically introduced in Grade 1, even those teachers should attend to whether the introduction of symbolism seems to assist the students in communication (which is the purpose of symbols) or whether it actually inhibits them. If it is the latter situation, then formal symbolism could be delayed.

Addition and Subtraction Language

You should model the following language when you work with addition and subtraction situations, although you should only expect students to use the terms *sum* and *difference* and only when they find the terms useful:

- The numbers you add are called the *addends* and the result is called the *sum*.
- When you subtract one number from another number, the number you subtract from is called the *minuend*, the number you subtract is called the *subtrahend*, and the result is called the *difference*. Be aware that the term *difference* is a tricky one. If we ask for the difference between 10 and 7, students are much more likely to talk about how those two numbers visually appear different (e.g., 10 has a 0 in it) than the result of a subtraction.

It is also important to attempt to read equations meaningfully. For example, in reading 12 − 5 = ❑, you might write the equation as shown, but actually say, *If I subtract 5 from 12, what is the result?* In reading 3 + ❑ = 8, you might write the equation, but say, *What do I need to add to 3 to get 8?* or *If 8 is made up of two parts, and one part has 3, how many are in the other part?*

Meanings of Addition

Although addition always relates to combining items, students find it easier first to consider active combination situations, where a joining actually occurs, before they move on to more static combination situations, where a whole is made up of two parts. (See the chart below.) It is important to model both meanings of addition for students and to expect them to create story problems using both meanings. Frequently, teachers focus too much on the joining meaning. It is particularly important to focus on part-part-whole combinations for ten since this makes adding with greater numbers much easier for students.

Teaching Idea 2.26

Prepare slips of paper with various addition expressions on them, such as 5 + 8 and 3 + 2. Have pairs of students choose a slip of paper and write two different story problems for the addition. Suggest that they make their two stories as different as possible.

To focus on BIWNO 2, ask: *How do you know that both your story problems are about addition even though they are different?* [Both problems match the same addition sentence and, in both problems, two things are put together.] *What is different about your story problems?* [e.g., In one problem, both parts were already there. In the other problem, one part was already there and the other part was added to it.]

Meanings of Addition as Applied to 5 + 3

MEANING	EXAMPLE
Joining A joining situation is an active combining situation.	A child has 5 marbles. His mom gives him 3 new marbles. 5 marbles + 3 new marbles = 8 marbles
Part-Part-Whole A part-part-whole situation is a static combining situation where no action takes place.	A child has 8 marbles. 5 marbles are blue and 3 marbles are red. 5 blue marbles + 3 red marbles = 8 marbles

Meanings of Subtraction

Subtraction is a more complex operation than addition. Most simply, it is the opposite of addition, but there are many nuances, or meanings. In the examples in the chart below, the same subtraction, 12 − 7 = 5, can be applied to each situation even though the situations appear different. Mathematically, they are the same.

It is important for students to realize that these different meanings for subtraction are related. In other words, it makes sense to apply the same operation—subtraction—in each situation, even though both the contexts and the problem situations are different.

Meanings of Subtraction as Applied to 12 − 7

MEANING	EXAMPLE
Take Away Taking away is an active separating situation.	 12 take away 7 leaves 5. 12 − 7 = 5
Comparison Comparing two quantities involves subtracting one quantity from the other.	 12 is 5 more than 7. 12 − 7 = 5
Missing Addend Finding a missing addend involves finding out how much to add.	 7 and 5 more is 12. 7 + ■ = 12 → 12 − 7 = 5

How the Meanings of Subtraction Are Equivalent

It is important for students to realize that these varying meanings for subtraction are related. In other words, it makes sense to apply the same operation—subtraction—in each situation even though both the contexts and the problem situations are different. You must help students explore these connections.

Consider the situation modelled and described below. To compare the number in the top row (the minuend) with the number in the bottom row (the subtrahend), you figure out how many more are in the top row than in the bottom row. You can take away the 3 counters that match in each row (which is equivalent to taking away 3 from 8) and what is left (the difference) is how many more were in the top row. So to compare to determine how many more, you can take away.

Relating comparison to take away

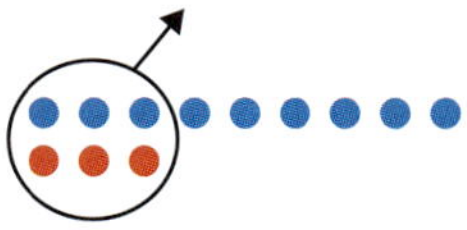

Anne has read 9 books. B.J. has read 3 books.
How many more books has Anne read than B.J.?
9 − 3 = 6

Teaching Idea 2.27

Tell students that you have 8 pens and 3 pencils. You want to know how many more pens you have than pencils. Then tell students that you had 8 pens but you lost 3 and you want to know how many are left. Have them model each problem with counters.

To focus on BIWNO 2, ask: *Why can you think about taking away 3 from 8 to figure out each answer?* [If you have a set of 8 pens and a set of 3 pencils and take away 3 items from each set, the 5 that are left shows how many more items were in the set of 8. Having 8 pens and losing 3 is starting with 8 and taking away 3.]

Understanding Addition and Subtraction Situations

The chart that follows summarizes the many different addition and subtraction situations students will encounter. Some involve actions, like joining or taking away (separating), some involve recognizing the parts that make up a whole, and some involve comparing. Sometimes the whole and one part are known and the other part must be determined; other times the two parts are known and the whole must be determined.

Each situation can be represented by either an addition or a subtraction number sentence with one number missing, and each is solved by either adding or subtracting the two known quantities. What is interesting is that an addition sentence can be solved sometimes by subtracting and sometimes by adding; the same is true of a subtraction sentence.

The most difficult situations for students are those represented by ■ + b = c and ■ − a = b, where the first term is unknown.

Addition and Subtraction Situations

	1	2	3
Joining Situations	**Result (sum) is unknown** *I had 3 pennies to start with. I got 1 more. How many do I have now?* 3 + 1 = ❑ *a* + *b* = ❑ To solve, you add.	**Change (addend) is unknown** *I had 5 pennies to start with. I got some more. Now I have 10. How many more did I get?* 5 + ❑ = 10 *a* + ❑ = *c* → *c* − *a* = ❑ To solve, you can subtract 10 − 5.	**Start (addend) is unknown** *I had some pennies to start with. I got 3 more. Now I have 8. How many did I start with?* ❑ + 3 = 8 ❑ + *b* = *c* → *c* − *b* = ❑ To solve, you can subtract 8 − 3.
	4	5	6
Separating (Take Away) Situations	**Result (difference) is unknown** *I had 6 pennies to start with. I spent 3. How many do I have left?* 6 − 3 = ❑ *c* − *a* = ❑ To solve, you subtract.	**Change (subtrahend) is unknown** *I had 6 pennies to start with. I spent some. I have 2 left. How many did I spend?* 6 − ❑ = 2 *c* − ❑ = *b* → *c* − *b* = ❑ To solve, you can subtract 6 − 2.	**Start (minuend) is unknown** *I had some pennies to start with. I spent 4. I have 6 left. How many did I start with?* ❑ − 4 = 6 ❑ − *a* = *b* → *b* + *a* = ❑ To solve, you can add 6 + 4.
	7	8	
Part-Part-Whole Situations	**Whole (sum) is unknown** *I have 3 pennies and 4 nickels. How many coins do I have altogether?* 3 + 4 = ❑ *a* + *b* = ❑ To solve, you add.	**Part (addend) is unknown** *I have 7 coins altogether. They are pennies and nickels. 3 are pennies. How many are nickels?* 3 + ❑ = 7 or ❑ + 3 = 7 *a* + ❑ = *c* → *c* − *a* = ❑ ❑ + *b* = *c* → *c* − *b* = ❑ To solve, you can subtract 7 − 3.	
	9	10	11
Comparing Situations	**Difference is unknown** *I have 7 pennies and 2 nickels. How many more pennies than nickels do I have?* 7 − 2 = ❑ *c* − *a* = ❑ To solve, you subtract.	**Comparing quantity (subtrahend) is unknown** *I have 7 pennies and some nickels. I have 2 more pennies than nickels. How many nickels do I have?* 7 − ❑ = 2 *c* − ❑ = *b* → *c* − *b* = ❑ To solve, you can subtract 7 − 2.	**Referent quantity (minuend) is unknown** *I have some pennies and 3 nickels. I have 2 fewer nickels than pennies. How many pennies do I have?* ❑ − 3 = 2 ❑ − *a* = *b* → *b* + *a* = ❑ To solve, you can add 2 + 3.

You should make sure students encounter a wide variety of these structures to help them construct a more complete understanding of addition and subtraction. The different meanings for the operations can be distinguished only through appropriate contexts and modelling.

The different meanings for the operations can be distinguished only through appropriate contexts and modelling.

Solving Addition and Subtraction Number Stories

Students must learn to decode and interpret number stories; they should not rely on looking for clue words to decide which operation to perform. Focusing on clue words is superficial and will only lead to errors; because of the relationship between the operations, the same clue word can describe both an addition and a subtraction situation. For example, in the Addition and Subtraction Situations chart, the word "more" appears both in Situation 1, which involves adding two numbers, and in Situation 2, which involves subtracting two numbers.

Students usually find joining and separating (take away) situations easiest to deal with. However, if you encourage students to model number stories, even young children can deal with part-part-whole and comparing situations.

Relating Addition and Subtraction

Notice that you can view any addition situation as a subtraction situation, and vice versa. A fact family is a set of two, three, or four number sentences, or equations, that describe the same situation. For example, you can use four different number sentences to describe the same set of 3 blue and 2 red squares, as shown below:

Fact Family for 2, 3, and 5

$3 + 2 = 5$	3 blue + 2 red is 5 altogether
$2 + 3 = 5$	2 red + 3 blue is 5 altogether
$5 - 2 = 3$	5 altogether; if 2 are red, then 3 must be blue
$5 - 3 = 2$	5 altogether; if 3 are blue, then 2 must be red

Teaching Idea 2.28

Ask students to think of a situation that $9 + 4 = 13$ might describe. [e.g., 9 baseballs and 4 footballs is 13 balls altogether]

Then, to focus on BIWNO 1, ask: *How can you describe the same situation using a subtraction?* [e.g., $13 - 4 = 9$. There are 13 balls altogether. 4 are footballs and you want to know how many other balls there are.]

Addition and Subtraction Principles

Students need to learn and discuss a number of principles about addition and subtraction.

ADDITION AND SUBTRACTION PRINCIPLES

1. Addition and subtraction "undo" each other. They are related inverse operations (e.g., if $4 + 8 = 12$, then $12 - 8 = 4$).
2. You can add numbers in any order (the commutative property). With subtraction, the order in which you subtract the numbers matters.
3. To add two numbers, you can take away from one number and add what you took away to the other number without changing the sum (the associative property).
4. You can add or subtract in parts (e.g., $2 + 5 = 2 + 4 + 1 = 7$; $8 - 5 = 8 - 3 - 2 = 3$).
5. To subtract two numbers, you can add or subtract the same amount to or from both numbers without changing the difference (e.g., $12 - 7 = (12 + 1) - (7 + 1)$).
6. When you add or subtract 0 to or from a number, the answer is the number you started with.
7. When you add 1 to a number, the sum is the next counting number. When you subtract 1 from a number, the difference is the counting number that comes before.

Some of these principles are restatements of what mathematicians call properties inherent in our number system, while others are built from those properties. Students do not need to memorize the principles by name, but they should become so familiar with our number system that they use the principles naturally and informally.

PRINCIPLE 1 Addition and subtraction "undo" each other. They are related inverse operations.

Mathematicians define subtraction as the inverse operation to addition; it undoes what addition accomplishes. Suppose you start with 4 items, add 8 more, and end up with 12. How do you get back to where you started at 4? You subtract the 8 you added. Similarly, if you start with 12 items and take away 4, you get back to the starting number, 12, by adding the 4 you took away.

PRINCIPLE 2 You can add numbers in any order (the commutative property). With subtraction, the order in which you subtract the numbers matters.

You can use counters to model and explain the commutative property, or order property. On a piece of paper, show a group of 3 counters and a group of 4 counters. Then turn the paper 180 degrees. It is clear that you have exactly the same counters, so $3 + 4 = 4 + 3$.

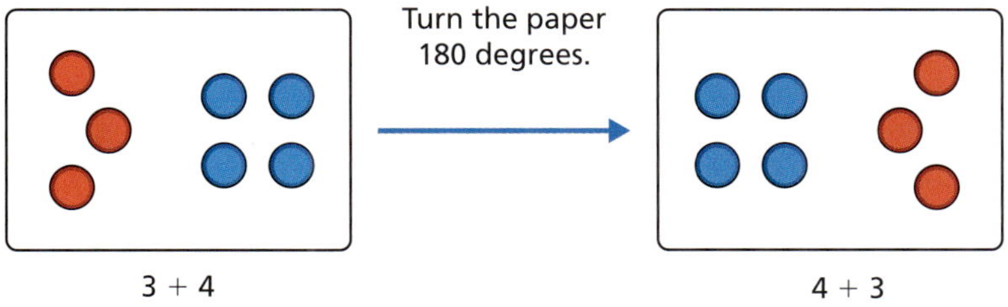

$3 + 4 = 4 + 3$

Modelling to explain the commutative property of addition

However, order does matter when you subtract. For example, if you take 3 cookies away from 7, you have 4 left. You cannot take 7 cookies away from 3, although students will learn later that you can use integers to calculate $3 - 7$.

PRINCIPLE 3 To add two numbers, you can take away from one number and add what you took away to the other number without changing the sum.

Suppose you are adding 4 counters and 6 counters, as shown below. You can move any number of counters from one pile to the other without changing the total. Moving 2 counters from the left pile to the right pile changes the calculation to $2 + 8$, but it does not change the sum. In fact, you have renamed $(2 + 2) + 6$ as $2 + (2 + 6)$.

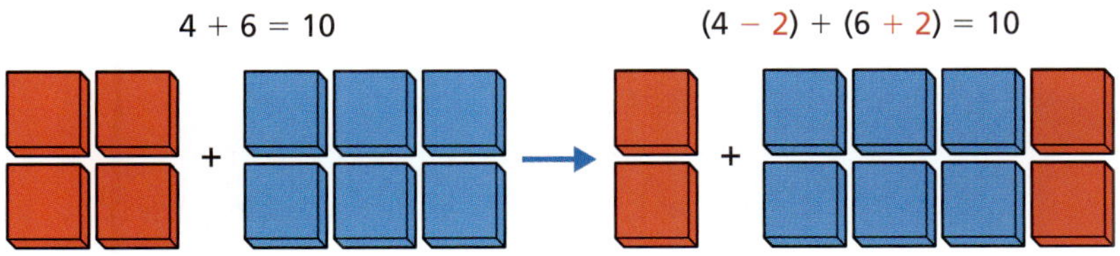

$4 + 6 = 2 + 8$

Modelling to explain the associative property of addition

This principle is based on the associative property that tells us that to add three numbers, you can add the first two numbers and then add the last

Teaching Idea 2.29

Have students tell how they would solve the following subtractions using addition:

$8 - 5$
$9 - 7$
$5 - 1$
$7 - 6$

To focus on BIWNO 1, ask students to respond to the statement: *There is no reason to learn to subtract because you can always add instead.* They can discuss the statement in pairs or small groups first.

number to the sum. Or, you can add the last two numbers and then add the first number to the sum. For example, to add $5 + 4 + 3$, you can add $5 + 4$ to get 9 and then add 3, or you can add $4 + 3$ to get 7 and then calculate $5 + 7$. The associative property had to be created because addition is defined in terms of how to combine two numbers; we needed a rule for combining three numbers.

A variation of Principle 3 is adding two numbers that are 2 apart by doubling the number in the middle, for example, $4 + 6 = 5 + 5$. If 1 is moved from 4 to 6, it is like doubling 5.

PRINCIPLE 4 You can add or subtract in parts.

Suppose you want to add 5 objects to a group of objects. It does not matter whether you add all 5 at once or you add them in stages. For example, if you add 3 and then add 2, you are still adding 5 ($2 + 5 = 2 + 3 + 2$). The same is true for subtraction. For example, to subtract 5 objects from a group, you can take away 3 and then take away 2 more ($8 - 5 = 8 - 3 - 2$).

PRINCIPLE 5 To subtract two numbers, you can add or subtract the same amount to or from both numbers without changing the difference.

If a boy is 12 years old and his brother is 7, the older brother is 5 years older than the younger brother. What about next year and the year after that? How much older will he be then? He will always be 5 years older because the difference remains unchanged. You can show this using a number model or you can model the situation using counters, as shown here.

$12 - 7 = 5$ $\quad\quad$ $(12 + 1) - (7 + 1) = 5$

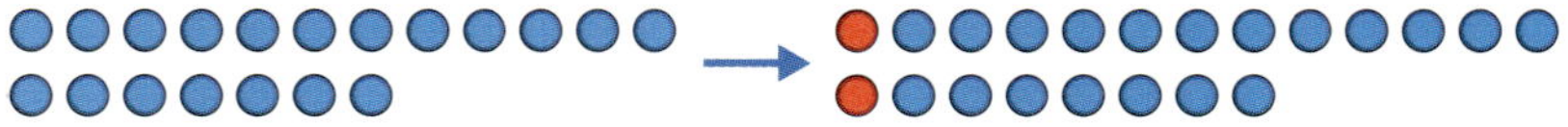

$12 - 7 = 13 - 8$

Modelling to explain how adding 1 to both numbers does not change the difference

PRINCIPLE 6 When you add or subtract 0 to or from a number, the answer is the number you started with.

A walk-on number line is a useful model to show this principle:

If $3 + 2$ means start at 3, step 2 spaces forward, and land on 5,

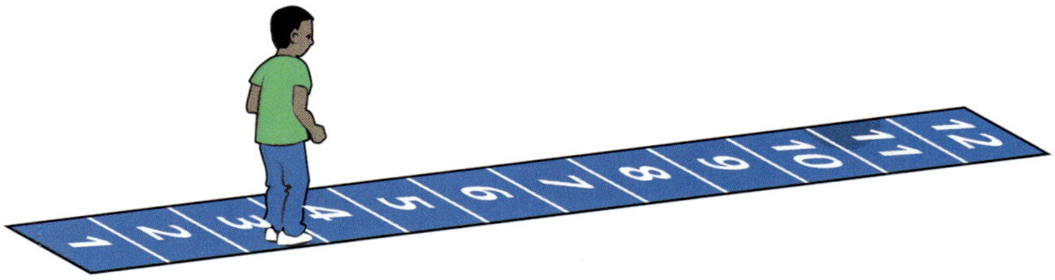

then $3 + 0$ means start at 3, step 0 spaces forward, and stay at 3.

PRINCIPLE 7 When you add 1 to a number, the sum is the next counting number. When you subtract 1 from a number, the difference is the counting number that comes before.

A walk-on number line is also helpful for demonstrating this principle, since adding 1 means to move 1 space forward, for example, $3 + 1 = 4$, and subtracting 1 means to move 1 space back, for example, $3 - 1 = 2$.

Teaching Idea 2.30

Ask students to think of three different ways to calculate $12 - 8$. [e.g., take away 8 from 10 and then add 2; take away 2 from 12 and then take away 6 more; count up from 8 to 12]

To focus on BIWNO 2, ask: *Is there always more than one way to subtract? How do you know?* [Yes; e.g., I can always count up from the second number to the first number (from 8 to 12 is 4) or I can count back from the second number to the first number (counting back from 12 to 8 is 4).]

Teaching Idea 2.31

Ask students why you could solve $16 - 9$ using $17 - 10$ or $18 - 11$ instead and talk about the pattern in the three subtractions. [You're adding 1 to each of the numbers but you get the same answer.]

To focus on BIWNO 2 and 3, ask: *What other subtraction would have the same result?* [e.g., $19 - 12$] *Why might someone choose to solve $16 - 9$ using $17 - 10$ instead?* [They have the same result, 7, but $17 - 10$ is easier to do mentally.]

Strategies for Learning Addition and Subtraction Facts

Students can apply the addition and subtraction principles to use known facts to find unknown facts.

Using the Principles Directly

There are 100 addition facts, from $0 + 0$ to $9 + 9$, as well as 55 related subtraction facts. (The reason there are fewer related subtraction facts is because subtraction does not have the commutative property, for example, $6 - 2$ is a subtraction fact but $2 - 6$ is not.) Students can use the addition and subtraction principles to help them use known facts to find unknown facts. For example,

Strategies for Relating Facts

PRINCIPLE USED	EXAMPLE
Principle 1 (subtraction and addition are related) You do not have to learn the subtraction facts as separate facts. To subtract, students can think of the related addition fact.	$8 + 3 = 11$, so $11 - 3 = 8$ and $11 - 8 = 3$
Principle 2 (you can add numbers in any order) You can relate 45 of the addition facts to 45 reverse facts, eliminating the need to memorize them separately.	$4 + 3 = 7$, so $3 + 4 = 7$

Using the Principles Indirectly

Many addition and subtraction facts are so closely related to known facts that they do not need to be learned separately.

Building on Doubles Students seem to learn the double facts easily, for example, $3 + 3 = 6$, $7 + 7 = 14$, and so on. They can use the principles to relate other facts to these facts.

Strategies for Using Doubles to Solve Related Addition Facts

PRINCIPLE(S) USED	EXAMPLE
Principle 3 (you can take away from one number and add what you took away to the other number without changing the sum)	$6 + 8$ *Move 1 from the 8 to the 6:* $6 + 8 = (6 + 1) + (8 - 1) = 7 + 7$ $7 + 7 = 14$
Principle 3, Principle 4 (you can add or subtract in parts) and **Principle 7** (when you add 1, the sum is the next counting number)	$7 + 8$ *Recognize that $8 = 7 + 1$:* $7 + 8 = 7 + (7 + 1) = 7 + 7 + 1$ $7 + 7 = 14$ and $14 + 1 = 15$

Making Tens Once students have learned the combinations for 10, for example, $4 + 6$, $7 + 3$, and so on, they can use the principles to relate other calculations to these. This is particularly efficient because once you can identify an addend as 10, you do not have to add to determine $10 + \square$ (if $\square$ is less than 10) because $10 + \square$ is a representation of the number $1\square$ (e.g., $10 + \underline{3} = 1\underline{3}$).

Strategies for Making Tens to Solve Addition Facts

PRINCIPLE USED	EXAMPLES	
Principle 3 (you can take away from one number and add what you took away to the other number without changing the sum)	$9 + 5$ *Move 1 from the 5 to the 9:* $9 + 5 = (9 + 1) + (5 - 1) = 10 + 4$ $10 + 4 = 14$	$8 + 6$ *Move 2 from the 6 to the 8:* $8 + 6 = (8 + 2) + (6 - 2) = 10 + 4$ $10 + 4 = 14$

You can also use the principles to relate subtraction facts to 10. Again, you do not need to use facts for $1\square - 10$ because $1\square - 10 = \square$ (e.g., $1\underline{3} - 10 = \underline{3}$).

Strategies for Making Tens to Solve Subtraction Facts

PRINCIPLE USED	EXAMPLES	
Principle 5 (you can add or subtract the same amount to or from both numbers without changing the difference)	$15 - 9$ *Add 1 to both numbers:* $15 - 9 = (15 + 1) - (9 + 1) = 16 - 10$ $16 - 10 = 6$	$15 - 8$ *Add 2 to both numbers:* $15 - 8 = (15 + 2) - (8 + 2) = 17 - 10$ $17 - 10 = 7$

Getting to 10 along the way is also a useful strategy for solving both addition and subtraction facts.

Strategies for Making Tens to Solve Facts

PRINCIPLE(S) USED	EXAMPLE
Principle 4 (you can add or subtract in parts)	$7 + \underline{6} = 7 + \underline{3 + 3} = 10 + 3 = 13$ $13 - \underline{4} = 13 - \underline{3 - 1} = 10 - 1 = 9$
Principle 1 (addition and subtraction are related) and **Principle 4**	$13 - 9 = \blacksquare \rightarrow 9 + \blacksquare = 13$ $9 + \underline{1} = 10$ and then $10 + \underline{3} = 13$ Since you add $1 + 3 = 4$ to get from 9 to 13, $13 - 9 = 4$

Teaching Idea 2.32

Ask students to describe an easy way to subtract 9 from any number. [e.g., Subtract 10 and then add back 1.]

To focus on BIWNO 4, ask: *Why is your method a good method?* [e.g., It's easier to subtract 10 and then add 1 than to subtract 9 because, to subtract 10, you just subtract 1 from the tens digit and to add 1, you just add 1 to the ones digit.]

It takes time to develop and nurture the thinking strategies that help students learn their addition and subtraction facts. Initially, you should emphasize the many relationships among the facts rather than the speed of recall. One of the best ways to apply and practise facts is in problem-solving situations like this:

You have 15¢ to spend at a yard sale. Pencils cost 2¢, erasers cost 3¢, and a pad of paper costs 5¢.

- *What items can you buy if you spend all of your money?* [e.g., 3 pads; 2 pads, 1 pencil, and 1 eraser; 5 erasers]
- *How much could three items cost?* [6¢, 7¢, 8¢, 9¢, 10¢, 11¢, 12¢, 13¢, 15¢]
- *If you spend only 13¢, what can you buy?* [1 pad and 4 pencils; 2 pads and 1 eraser; 3 erasers and 2 pencils]

Multiplication and Division

The other two important operations in the elementary grades are multiplication and division. As with addition and subtraction, a solid foundation in the various meanings and principles of multiplication and division is vital for developing students who are flexible and proficient at doing the calculations that are essential to problem solving.

Multiplication and Division in the Early Grades

In Kindergarten through Grade 2, teachers offer problems for students to solve that might be classified as multiplication and division problems, but without any formal multiplication or division language. Problems implicitly involve these operations when equal groups are involved. The problems may seem like counting problems or addition or subtraction problems to the students, not multiplication and division problems. An example might be a teacher suggesting that there were 4 plates, each with 3 cookies and asking, *How many cookies altogether are on the plates? If I moved all the cookies to just 2 plates, with the same number of cookies on each, how many cookies would be on each plate?*

Most young students will use some sort of model, such as counters, to help them solve the problem, although some students might use addition or subtraction thinking. For example, a child might think $3 + 3 + 3 + 3 = 12$. They might then push 12 counters into 2 piles to determine the size of each pile. It is only later on that formal symbolism and language are added to the situation, for example, a teacher asking if the students had multiplied or divided to determine each result.

Multiplication and Division Language

You should model the following language when you work with multiplication and division situations, although you should only expect students to use the terms *product* and *quotient* only as they are ready.

- When you multiply one number by another number, the numbers you multiply are called *factors* and the result is called the *product*.
- When you divide one number by another number, the number you divide is called the *dividend*, the other number is called the divisor, and the result is called the *quotient*.

As was mentioned earlier, the use of the more formal language should facilitate communication and not inhibit it. A teacher must make a judgment about when the introduction of this language is appropriate for particular students.

The use of the more formal language should facilitate communication and not inhibit it.

Meanings of Multiplication

At the K to 3 level, students focus only on some meanings of multiplication and division; more meanings are introduced in later grades (combinations, rate/ratio, area, and fractions). The usual meanings for multiplication taught in K to 3 are shown below. Early on, in Kindergarten and Grade 1, the main focus is likely to be on equal groups.

It is important that students learn to distinguish between an addition situation that can be modelled with multiplication (because all groups are the same size) and one that cannot (where not all groups are the same size).

Meanings of Multiplication as Applied to 3 × 4

MEANING	EXPLANATION AND EXAMPLE
Equal Groups or Sets	3×4 is the total number of objects in 3 sets of 4.  $4 + 4 + 4 = 12$ $3 \times 4 = 12$ This model shows both equal groups and repeated addition.
Repeated Addition	The first factor, in this case 3, tells how many times to add the second factor, 4. See the model above. $3 \times 4 = 4 + 4 + 4$
An Array	3×4 is the total number of items in a 3-by-4 array. 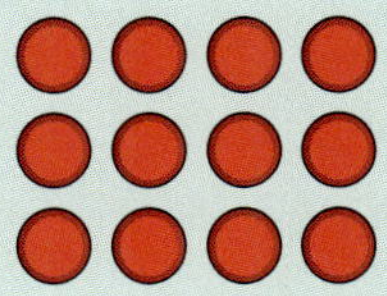 $3 \times 4 = 12$ An array of 3 rows with 4 counters in each row has 12 counters altogether.

Teaching Idea **2.33**

To focus on BIWNO 1 and 2, pose a problem where you show 4 tricycles and ask students how to use addition and then multiplication to determine the total number of wheels. [3 + 3 + 3 + 3 = 12, 4 × 3 = 12] Then show 4 tricycles and 1 bicycle and ask: *Why can you not use only multiplication to determine the total number of wheels?* [The groups of wheels are not the same size. I can multiply to find the number of wheels on the 4 tricycles but I have to add 2 to include the bicycle's wheels.] Ask students to create two more story problems, one that can be solved using only multiplication and one that cannot.

How the Meanings of Multiplication Are Equivalent

It is important for students to understand why the various meanings of multiplication are equivalent so that they can recognize a variety of situations to which multiplication can be applied. One such relationship between the meanings is shown below.

You can determine the number of items in a 3-by-4 array using repeated addition by adding the number in each row, $4 + 4 + 4 = 12$. You can also think of it as 3 equal groups of 4, where the counters in each row make up a group, $3 \times 4 = 12$.

Relating repeated addition, equal groups, and an array

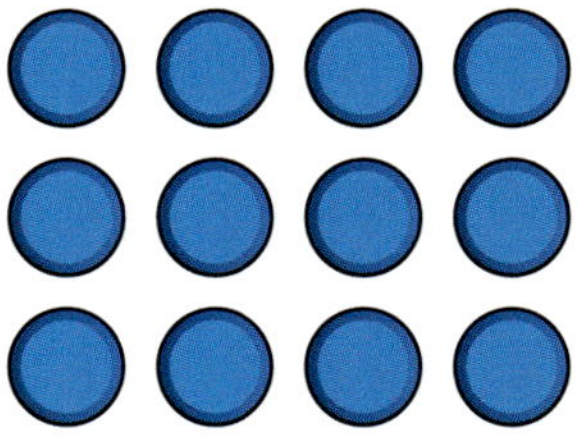

3 rows of 4 is 3 groups of 4
$4 + 4 + 4 = 3 \times 4$

Meanings of Division

At the K to 3 level, students are often exposed to three meanings of division.

Meanings of Division as Applied to 12 ÷ 4

MEANING	EXPLANATION AND EXAMPLE
Equal Sharing (Partitive)	12 ÷ 4 = 3 is the amount each person gets if you share 12 items equally among 4 people. 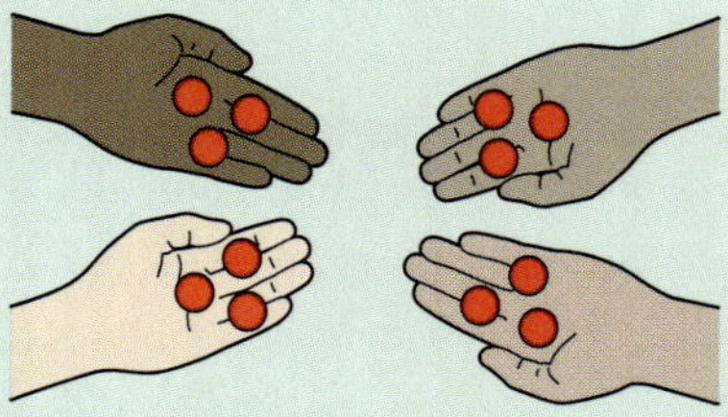 12 ÷ 4 = 3
Equal Grouping (Quotative)	12 ÷ 4 = 3 is the number of equal groups of 4 you can make with 12 items. 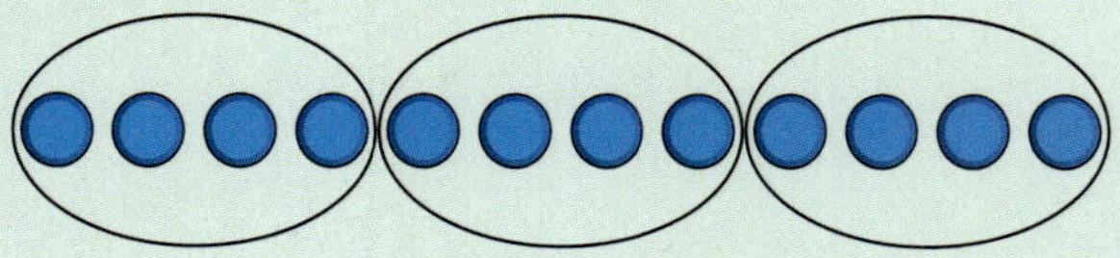 12 ÷ 4 = 3
Repeated Subtraction	12 ÷ 4 = 3 is the number of times you can subtract 4 from 12 before you get to 0. 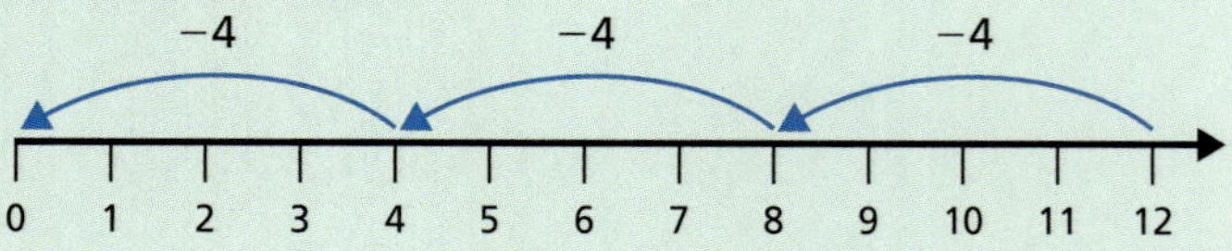 12 − 4 − 4 − 4 = 0 12 ÷ 4 = 3

Teaching Idea **2.34**

Ask students to tell what 15 ÷ 3 might mean. [e.g., how many people can share 15 items if each person gets 3 items]

To focus on BIWNO 1, ask: *Someone said you can figure out the answer by multiplying instead of dividing. Do you agree? Why?* [Yes; Figuring out how many groups of 3 are in 15 is like figuring out what to multiply 3 by to get 15.]

At the Kindergarten and Grade 1 levels, students might solve problems involving either sharing or equal grouping, although no formal division symbolism or language is used. Some sample problems are

- *Mom only gives me two cookies for lunch each day. How many days will 12 cookies last?*
- *I have ten stickers. If I share them with you fairly, how many will each of us get?*

How the Meanings of Division Are Equivalent

It is important for students to understand why the various meanings of division are equivalent. One such relationship is shown below.

Suppose you want to share 12 items among 4 people. One of the most natural ways to do this is to give each of the 4 people 1 item, then give each of them a second item, and then a third item, until all the items are gone. When you have given each of the 4 people his or her first item, you have created a group of 4. When you have given each of the 4 people a second item, you have created another group of 4, and so on. So, even though you are sharing, you are also creating equal groups with each round of sharing.

Relating sharing to creating equal groups

Sharing 12 cookies among 4—with each round of sharing, you create an equal group.

Teaching Idea | **2.35**

Pose this problem:

Rebecca had $24. She bought some headbands that cost $4 each. She had $12 left.

To focus on BIWNO 1, ask: *How can you use subtraction to figure out how many headbands she bought?* [Figure out how many 4s to subtract from 24 to get to 12.] *How can you use division?* [Divide 12 by 4.] *How can you use multiplication?* [Figure out what to multiply 4 by to get 12.]

Understanding Multiplication and Division Situations

As with addition and subtraction, there are many types of multiplication and division situations, only some of which students normally meet in K to 3. Each situation can be represented by either a multiplication or division number sentence or equation with one number missing and can be solved by either multiplying or dividing the two known quantities. Sometimes you can solve a multiplication by dividing and sometimes you can solve a division by multiplying.

The most difficult situations for students are those represented by ■ × b = c and ■ ÷ a = b, where the first term is unknown.

Multiplication and Division Situations

1	2	3
Whole (product) is unknown *I have 3 boxes with 4 tomatoes in each box. How many tomatoes are there?* 3 × 4 = ■ *a* × *b* = ■ To solve, you multiply.	**Size of group (quotient) is unknown** *I have 12 tomatoes packed equally in 3 boxes. How many tomatoes are in each box?* 12 ÷ 3 = ■ → 3 × ■ = 12 *c* ÷ *a* = ■ → *a* × ■ = *c* To solve, you can divide (but likely by relating it to a multiplication).	**Number of groups (divisor) is unknown** *I have 12 tomatoes in some boxes. There are 4 tomatoes in each box. How many boxes are there?* 12 ÷ ■ = 4 → ■ × 4 = 12 *c* ÷ ■ = *b* → ■ × *b* = *c* To solve, you can divide (but likely by relating it to a multiplication).

As with addition and subtraction, you should make sure students encounter a variety of multiplication and division situations. Once students are comfortable solving simple multiplication and division number stories, they may wish to create their own number stories.

Teaching Idea 2.36

Ask students to draw a picture to show why $6 \times 3 = 18$. Then ask what division the picture shows.

To focus on BIWNO 1, ask: *Can you always see a division in a multiplication picture? Why?* [Yes; When I show equal groups of items for multiplying, I can divide the total number of items by the number of groups (or number in each group) to show division.]

Relating Multiplication and Division

Notice that you can view any multiplication situation as a division situation and vice versa. In fact, what teachers call a fact family is a set of two to four number sentences, or equations, that describe the same situation.

Fact Family for 4, 3, and 12

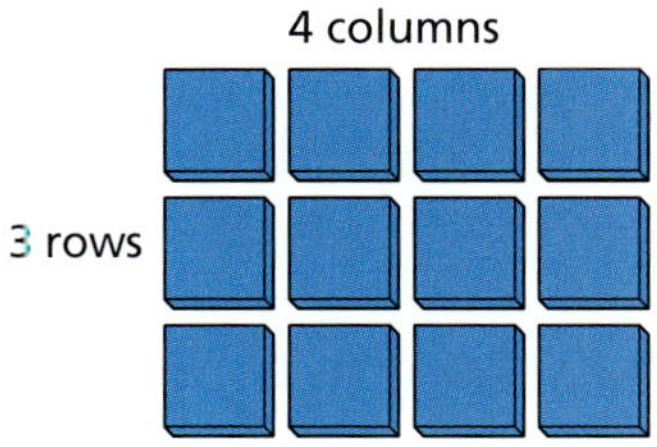

$4 \times 3 = 12$	4 columns of 3 squares are 12 squares
$3 \times 4 = 12$	3 rows of 4 squares are 12 squares
$12 \div 4 = 3$	12 squares in 4 columns form 3 rows
$12 \div 3 = 4$	12 squares in 3 rows form 4 columns

At the K to 3 level, you should have students focus on the meanings of multiplication and division and on solving simple problems involving those operations rather than on performing algorithms or using more sophisticated fact strategies.

Some students are likely to memorize some multiplication facts in the course of solving problems, but that should not be the focus of instruction.

Addition and Subtraction Procedures

Estimating Sums and Differences

It is valuable to estimate sums and differences because it helps us check calculations and predict answers. To solve some problems, you only need an estimate. This is equally true when adding or subtracting 2-digit, 3-digit, or greater numbers.

Teaching Idea 2.37

Tell students that you are selling one item for \$48 and another item for \$58. Ask whether \$100 is enough to pay for both items. Ask how they know.

To focus on BIWNO 5, ask: *Would estimating \$48 + \$58 by rounding both numbers down to 40 + 50 be a good way to see if \$100 is enough? Why or why not?* [No; When I'm estimating to see if \$100 is enough, a high estimate is safer. If I round both numbers down, I have a low estimate.] *How would you estimate instead?* [Round both numbers up to 50 + 60.]

Factors that Influence Estimation

A number of factors come into play when we make decisions about estimating, such as the context and the numbers and operations involved.

Context The context determines

- whether an estimate or an exact answer is more appropriate. For example, when a customer wants to know whether she or he can afford two items, an estimate might be appropriate. But when the clerk is giving the customer change, an exact difference is required.
- how close the estimate should be to the exact value. For example, estimating the total number of people at a sports event for a news report does not require as close an estimate as estimating the total cost of two items.
- whether a high or low estimate is more appropriate. For example, to estimate the total cost of two items, a high estimate is safer than a low estimate. A low estimate is probably better when someone is estimating how many of something he or she can afford to buy.

The Numbers and Operations Involved

- The individual numbers involved matter when you decide how to estimate. For example,
  - 836 + 94 is about 836 + 100 = 936, rounding 94 to 100 because 100 is close and easy to work with.
  - 672 − 69 is about 675 − 75 = 600, rounding both numbers to multiples of 25 that are compatible because multiples of 25 are easy to work with.
- Whether you are adding or subtracting also matters:
  - Rounding one number up and the other down is usually appropriate for adding, but not necessarily for subtracting, for example, 46 + 36 is about 50 + 30 (rather than 50 + 40).
  - Rounding both numbers up or both numbers down is usually appropriate when subtracting, but not when adding, for example, 84 − 36 is about 90 − 40 (rather than 80 − 40).

Teaching Idea 2.38

Have students estimate each and describe their strategies:

49 + 37   49 − 37

To focus on BIWNO 5, ask: *Why might someone round 37 down to estimate 49 + 37 as 50 + 35 but round 37 up to estimate 49 − 37 as 50 − 40?* [When you estimate a sum, it's better to round one number up and the other down because it's like moving some of the 37 to the 49. When you estimate a difference, it's better to round both numbers up or both down because the difference doesn't change if you change both numbers the same way.]

Estimation Strategies

To estimate sums and differences, students might use strategies like these:

ESTIMATION STRATEGY	EXAMPLE(S)
Round numbers to the nearest multiple of 10, 100, 1000, ...	693 + 458 is about 700 + 500 = 1200 693 − 458 is about 690 − 460 = 230
Round numbers to another "round" number.	693 + 458 is about 700 + 450 = 1150 693 − 458 is about 700 − 450 = 250
Round only one of the numbers.	693 + 458 is about 700 + 458 = 1158
When adding, round one number up and one down.	693 + 458 is about 700 + 450 = 1150
When subtracting, round both numbers up or down.	693 − 458 is about 700 − 500 = 200 693 − 458 is about 650 − 450 = 200
Estimate within a range.	428 + 397 is between 700 (400 + 300) and 900 (500 + 400) 516 − 147 is between 300 (500 − 200) and 500 (600 − 100)
Think in terms of money.	385 + 245 is about 3 loonies, 3 quarters + 2 loonies, 2 quarters, which is 6 loonies, 1 quarter, or 625
Use compatible numbers.	867 − 471 is about 867 − 467 = 400

Varied Approaches to Addition and Subtraction

As early as Grade 2, students are expected to apply what they know about adding and subtracting single-digit numbers ("the facts") and the meanings of those operations to adding and subtracting greater numbers. At this point it is useful to learn algorithms or procedures they can follow to add and subtract greater numbers. Before students can work efficiently and flexibly with algorithms, they must

- know their addition and subtraction facts
- understand the basic Numeration Principles that underlie the place value system
- know how to add and subtract multiples of 10, 100, 1000, ...
- understand the basic Addition and Subtraction Principles, because many algorithms are built on those principles

The transition from invented to conventional algorithms is best done when a student is ready, not necessarily according to a teacher's timetable.

Invented Algorithms

You should give students early opportunities to develop their own algorithms. These are sometimes referred to as invented or personal algorithms. Research has shown that students have greater success if they have opportunities early on to develop their own algorithms, generally using concrete models (Hiebert, 1984; Kamii, Lewis, and Livingston, 1993).

Teaching Idea 2.39

Ask students to tell you three different ways to add 42 to 89. [e.g., start with 42, add 100, then subtract 10 and subtract 1; add 41 and 90; add 50 to 81]

To focus on BIWNO 4, ask: *Which way makes the most sense to you? Why?* [e.g., add 100, subtract 10, and subtract 1; Each of those subtractions is really easy to do in my head.]

Mental Math

Some of the algorithms presented on the next few pages are more appropriate for pencil-and-paper work, supported initially with concrete or pictorial representations. Others can be mental algorithms, depending on the numbers involved and the abilities of students.

There is great value in using mental algorithms, as they are often quick and easy. It is often helpful to introduce mental algorithms with visual support, whether a number line, a beaded number line, or a 100 chart.

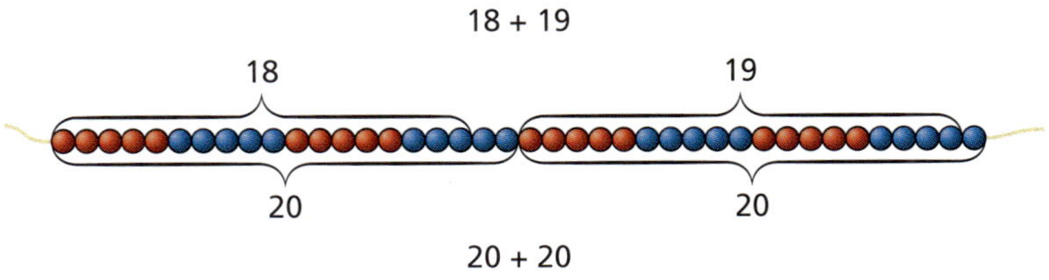

Using a beaded number line to show why 18 + 19 = 20 + 20 − 2 − 1

Students who do mental calculations apply and further develop their number sense.

Most mental algorithms require students to use Addition and Subtraction Principles 3, 4, and 5 to compose and decompose numbers to create different representations. For instance, they can think of 99 + 36 as 100 + 35, or think of 111 − 89 as 100 − 90 + 10 + 2. Mental computation is about considering which representation of a number is most useful for a particular calculation.

One very effective way to encourage students to calculate mentally is to present computations in horizontal form, rather than vertically. In this way, students do not immediately think in columns of digits. This helps them to remember that, for example, the 2 in 324 is 20 and not 2.

You should model appropriate language and expect students to use it.

Communicating about Algorithms

Both you and your students should use the terms *regroup* and *trade* when you talk about addition and subtraction. Do not use *carry* for addition or *borrow* for subtraction. Carrying and borrowing have no real meaning with respect to the operation you are performing, but the term *regroup* describes the action you must take. In the example below, when the tens are added, the result is more than 9 tens, so the 12 tens are regrouped into 1 hundred, 2 tens ("carrying 1" to the hundreds place does not describe this action meaningfully).

Students and teachers sometimes use poor math language when performing algorithms that focus on one place value or digit at a time. In the middle column of the example below, you are not adding 9 and 3. Rather, you are adding 9 tens and 3 tens, or 90 and 30.

ADD THE ONES	ADD THE TENS AND REGROUP	ADD THE HUNDREDS
395 + 134 9 5 + 4 = 9	1 395 + 134 29 9 tens + 3 tens = 12 tens 12 tens = 1 hundred, 2 tens	1 395 + 134 529 3 hundreds + 1 hundred + 1 hundred = 5 hundreds

Alternative Algorithms

Many people are surprised to learn that there is considerable flexibility in calculating sums and differences, just as there is with estimating.

Why Alternative Algorithms Are Important

Teachers should model and encourage alternative algorithms because

- One algorithm might make more sense to one student than to another student.
- One algorithm might be more convenient for a particular set of numbers.
- Some algorithms lend themselves to mental computation.
- Students may get help from parents who use different algorithms than those taught at school. It is helpful if students are open to both.
- Students who have a repertoire of algorithms can use one algorithm to perform a calculation and a different algorithm to check it.
- Some algorithms are procedures that students might invent. Something a student creates himself or herself is often more meaningful, and sometimes more efficient than a conventional algorithm.

Teachers are sometimes disappointed when a student uses the traditional algorithm to calculate a computation such as 1000 − 999 rather than using an alternative algorithm that reflects a more common-sense approach, as shown below. This is hardly surprising when you consider that the traditional algorithm is sometimes the only algorithm taught and that students are sometimes penalized for not using the traditional algorithm.

Teaching Idea 2.40

Ask students to use the digits 4, 7, 5, and 8 to create a subtraction question. Then ask what addition or subtraction facts they would need to know to calculate the difference. [e.g., for 85 − 47 they need 15 − 7 and 7 − 4]

To focus on BIWNO 4, ask: *Would it be possible to calculate [85 − 47] if you did not know [15 − 7]? Explain.* [Yes; e.g., I can solve 85 − 47 by adding up from 47 to 85, which is 3 from 47 to 50 and another 35 to 85.]

Subtracting 1000 − 999

USING THE TRADITIONAL ALGORITHM	USING AN ALTERNATIVE ALGORITHM
The traditional algorithm focuses on single digits: 9 9 10 ~~1000~~ − 999 1	This alternative algorithm uses number sense and addition principles. It lends itself to mental math: 1000 − 999 = ■ → 999 + ■ = 1000 999 + 1 = 1000 1000 − 999 = 1

Alternative Addition Algorithms

In the examples below, the algorithms apply equally well to 2-digit and 3-digit addition. For 2-digit sums, other materials such as Unifix trains of tens and single cubes, or bundles of ten popsicle sticks and single sticks would be a likely alternative to base ten blocks. For both 2-digit and 3-digit sums, money models (dimes and pennies or loonies, dimes, and pennies) can be used if students are comfortable with non-proportional representations, as can counters on place value charts.

For the traditional algorithm (Algorithm 1 shown on the next page) and many others, it is appropriate first to model the algorithm with manipulatives. In the early stages of using an algorithm, a written record is not necessary, but later on it is important to record each step of the algorithm to match a physical action with the manipulatives. No matter which algorithm is used, it is important that students understand and are able to explain why they do what they do. Algorithms suitable for mental calculation are noted as they are described.

For 2-digit sums, place value materials such as trains of ten and single cubes, or bundles of ten popsicle sticks and single sticks might be used instead of base ten blocks.

To use the traditional algorithm, you combine the digits, beginning at the ones place and work from the right to the left.

Addition Algorithm 1

Step 1 Model both numbers with blocks on a place value mat, if available.

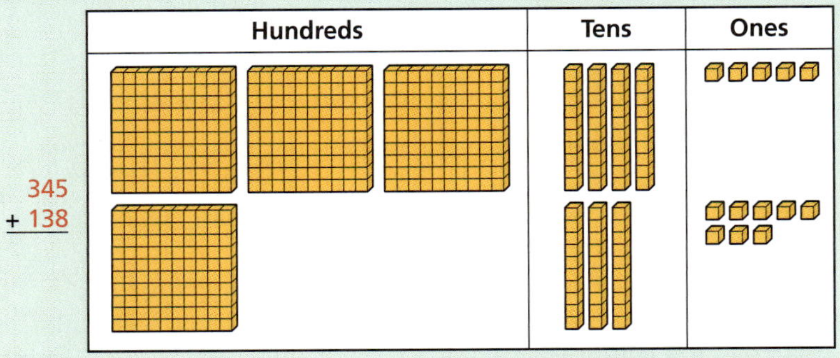

Step 2 Add the ones. Trade 10 ones for 1 ten. Record the ones that are left.

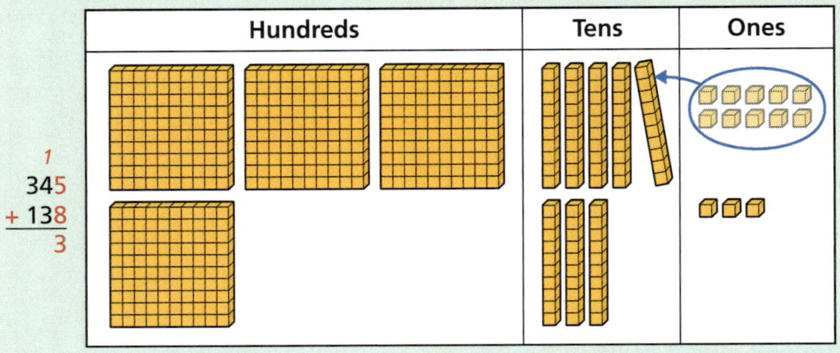

Step 3 Add the tens. Record the tens.

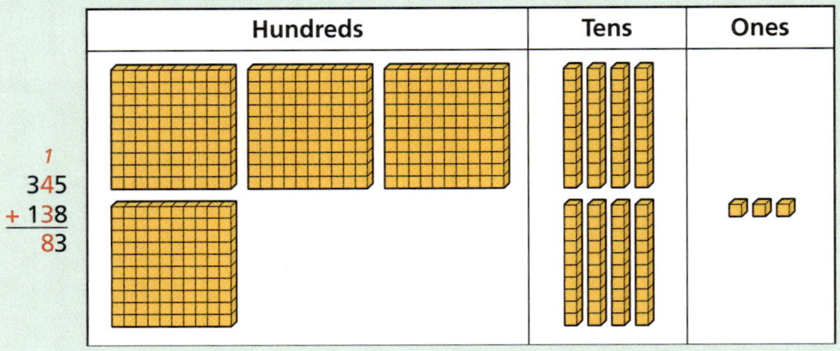

Step 4 Add the hundreds. Record the hundreds.

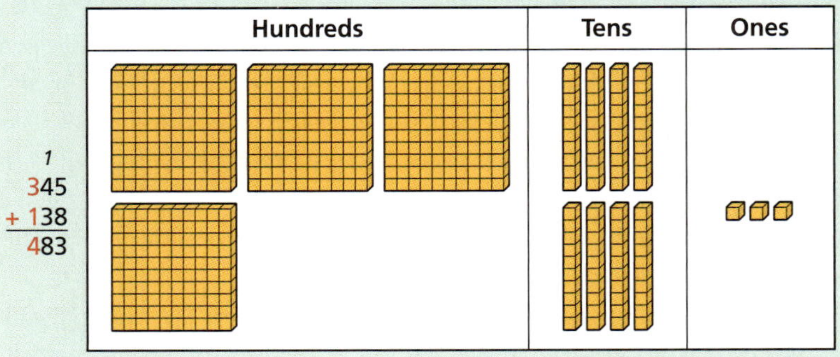

You might choose to have students work on a place value mat with three rows: one for the first addend, one for the second addend, and one for the sum. If there are enough blocks, students can even leave the addend blocks in place and use other blocks to show the sum. In this way, there is a complete concrete record of the question.

Addition Algorithm 2

STARTING AT THE LEFT		STARTING AT THE RIGHT	
345		345	
+ 138		+ 138	
400	Add the hundreds.	13	Add the ones.
70	Add the tens.	70	Add the tens.
+ 13	Add the ones.	+ 400	Add the hundreds.
483	Add the partial sums.	483	Add the partial sums.

This algorithm involves determining partial sums and then totalling them. It lends itself to mental math. You can start from the left or from the right, although starting from the left is better for mental math.

Modelling Algorithm 2 with Base Ten Blocks

This models the version of Algorithm 2 that starts at the left.

Step 1 Model both numbers with blocks using a place value mat, if available.

345
+ 138

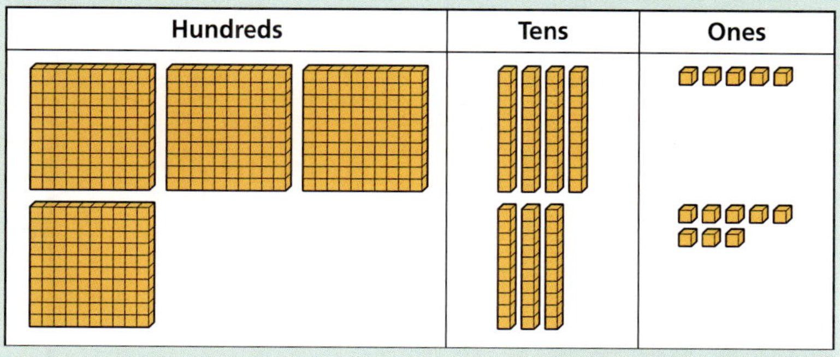

Step 2 Add the hundreds. Record the partial sum.

When you use this version of the algorithm, by Step 2 you have a low estimate of the sum.

345
+ 138
400

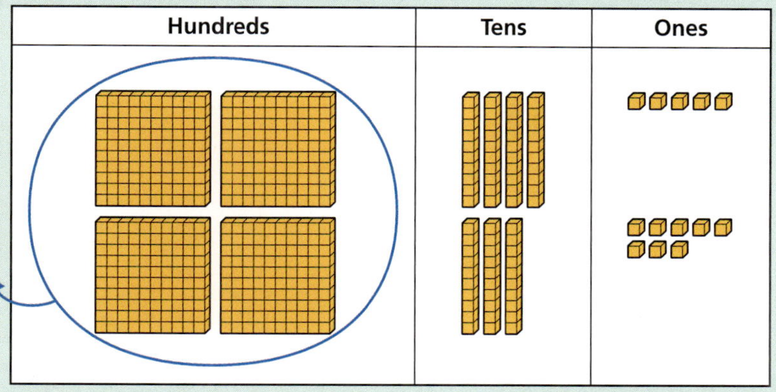

Step 3 Add the tens. Record the partial sum.

345
+ 138
400
70

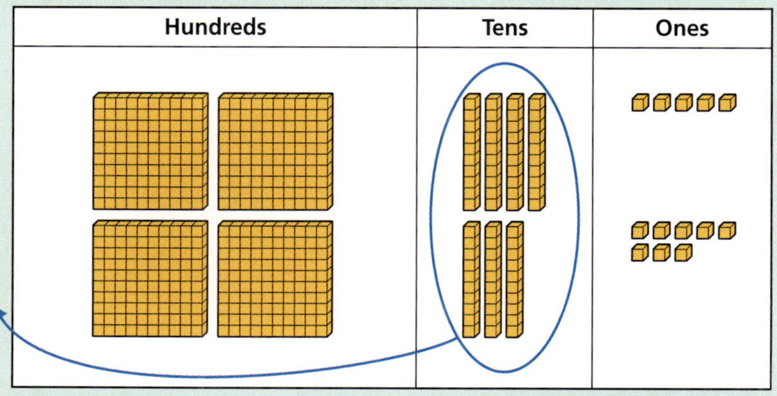

(continued)

Modelling Algorithm 2 with Base Ten Blocks (continued)

Step 4 Add the ones. Record the partial sum.

$$\begin{array}{r} 345 \\ +\,138 \\ \hline 400 \\ 70 \\ 13 \end{array}$$

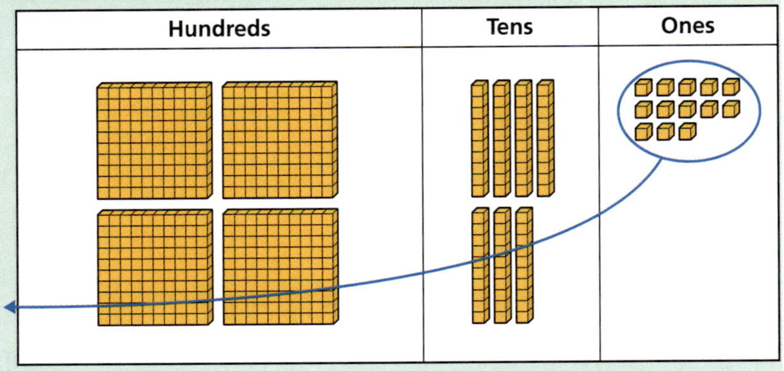

Step 5 Combine the hundreds, tens, and ones. Regroup 10 ones for 1 ten. Record the sum of the partial sums.

$$\begin{array}{r} 345 \\ +\,138 \\ \hline 400 \\ 70 \\ +\,13 \\ \hline 483 \end{array}$$

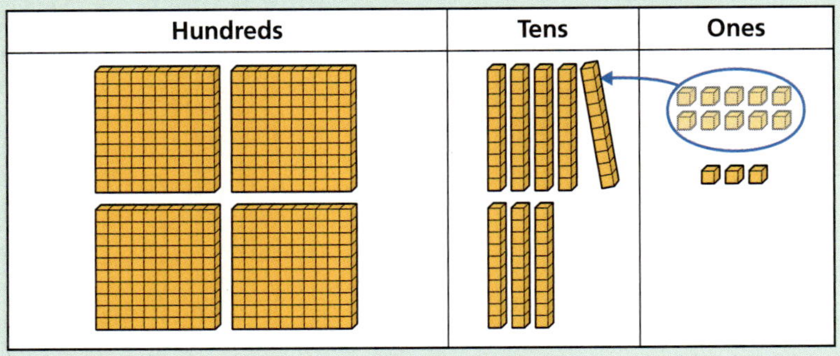

Teaching Idea 2.41

Ask students why you might solve 385 + 397 by adding 385 + 400 and then subtracting 3. [Adding 400 and subtracting 3 is the same as adding 397 because 400 − 3 = 397, and it's easier to add hundreds and then count back a small number than to add a number like 397.]

To focus on BIWNO 2, ask: *List three other additions where you might add too much and then subtract to make the addition easier to do. Show how you would do it.* [e.g., 47 + 49 = 50 + 50 − 4; 24 + 88 = 25 + 90 − 3; 187 + 412 = 200 + 412 − 13]

Teaching Idea 2.42

Ask students to solve these two additions:

501 + 498   536 + 214

To focus on BIWNO 4, ask: *Why might you do these two additions differently?* [e.g., For 501 + 498, it's easy to change it to 500 + 499, but for 536 + 214, it's easier to add 500 + 200 and then add 36 + 14.]

Addition Algorithm 3

This algorithm is based on **Principle 1** (addition and subtraction "undo" each other). In this case, you add too much in order to get an addend that is easier to work with, often mentally, and then undo what you did by taking away.	35 + 59 *Add 1 too many:* 35 + 59 → 35 + 60 = 95 *Take away the extra 1:* 95 − 1 = 94 35 + 59 = 94	357 + 597 *Add 6 too many:* 357 + 597 → 360 + 600 = 960 *Take away the extra 6:* 960 − 6 = 954 357 + 597 = 954

Addition Algorithm 4

This algorithm is based on **Principle 3** (you can take away from one number and add what you took away to the other number without changing the sum). You create addends that are easier to calculate, often mentally.	34 + 98 *Add 2 to one addend and subtract 2 from the other:* 34 + 98 = (34 − 2) + (98 + 2) = 32 + 100 = 132 34 + 98 = 132	389 + 197 *Add 3 to one addend and subtract 3 from the other:* 389 + 197 = (389 − 3) + (197 + 3) = 386 + 200 = 586 389 + 197 = 586

Addition Algorithm 5

This algorithm is based on **Principle 4** (you can add in parts).	47 + 56 *Add 56 in parts (3 + 53):* *First add 3:* 47 + 3 = 50 *Then add 53:* 50 + 53 = 103 47 + 56 = 103
	345 + 138 *Add 138 in parts (100 + 30 + 8):* *First add 100:* 345 + 100 = 445 *Then add 30:* 445 + 30 = 475 *Then add 8:* 475 + 8 = 483 345 + 138 = 483

Note that it is easiest to use Algorithm 5 when adding subsequent amounts does not affect previous additions (as shown above). For a calculation such as 345 + 178, you need to regroup the tens digits when you add 70 (of 178) to 40 (of 345).

You can often perform Algorithm 5 mentally.

Alternative Subtraction Algorithms

As with addition, there is a broad range of subtraction algorithms. A number of these are described here. Algorithms suitable for mental calculation are noted.

These algorithms apply to 2-digit and 3-digit subtraction and alternative place value materials might be used instead of base ten blocks.

The traditional North American algorithm (Algorithm 1 shown below) is not always the easiest for students. One of the reasons is that it proceeds in a left-to-right direction to regroup but in a right-to-left direction to subtract. The algorithm focuses on one place value position at a time and consequently can involve several stages of regrouping. At each place value, students first have to decide whether to regroup, and then perform the regrouping. This requires students to go back and forth between thinking about regrouping and performing the subtraction.

Subtraction Algorithm 1

Step 1 Model the minuend on a place value mat, if available.

312
− 178

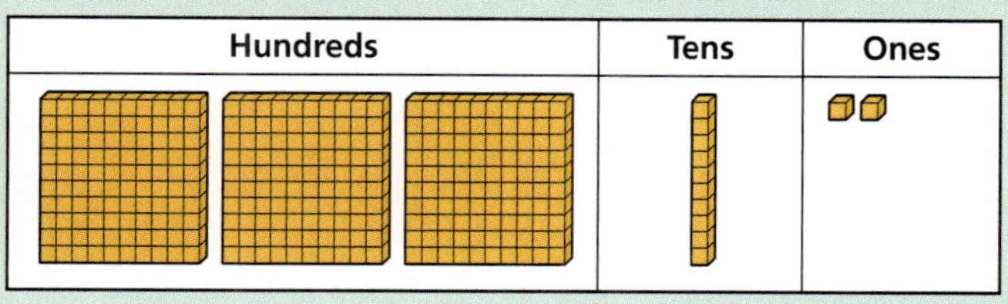

You do not model the subtrahend in a take-away subtraction situation because it is to be taken away from the minuend.

Step 2 Recognize that you do not have enough ones to be able to take away 8 ones. Trade 1 ten for 10 ones so you can take away 8 ones. Record the ones that are left.

0 12
3~~12~~
− 178

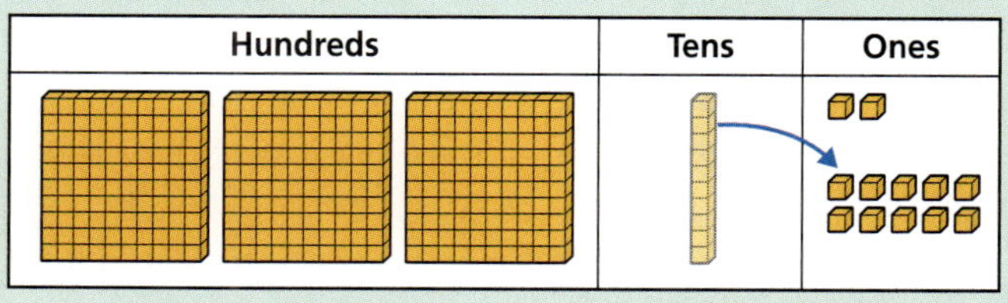

0 12
3~~12~~
− 178
4

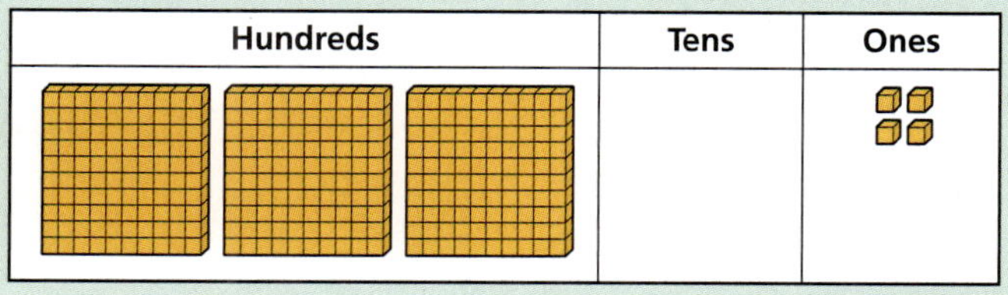

(continued)

For 2-digit subtraction, place value materials such as trains of ten and single cubes, or bundles of ten popsicle sticks and single sticks might be used instead of base ten blocks.

This algorithm proceeds in a left-to-right direction to regroup but in a right-to-left direction to subtract, requiring students to go back and forth between thinking about regrouping and performing the subtraction.

Subtraction Algorithm 1 (continued)

Step 3 Recognize that you do not have enough tens to be able to take away 7 tens. Trade 1 hundred for 10 tens so you can take away 7 tens. Record the tens that are left.

2 10 12
~~3~~~~1~~2
− 178
4

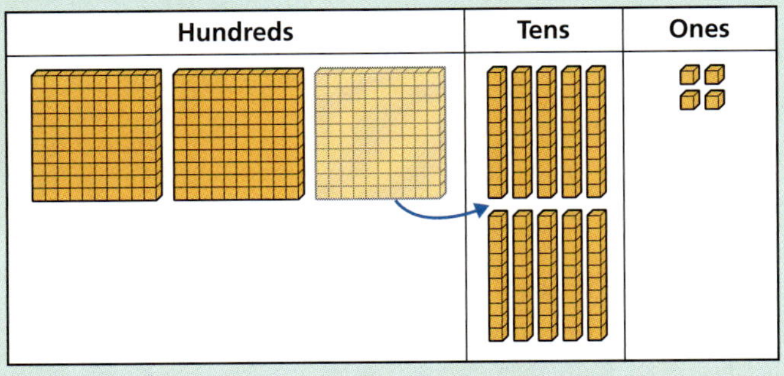

2 10 12
~~3~~~~1~~2
− 178
34

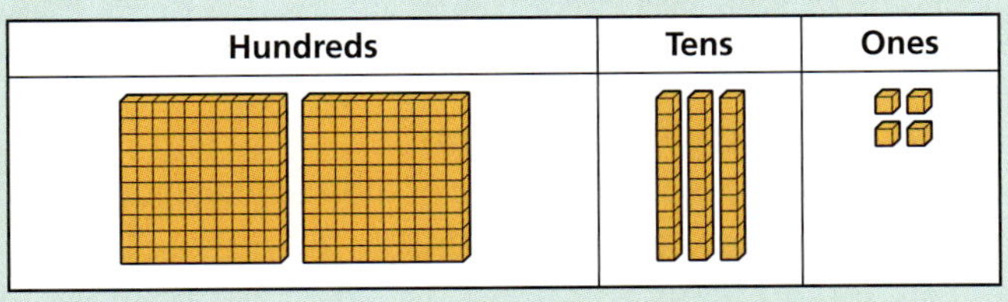

Step 4 Take away 1 hundred. Record the hundreds that are left.

2 10 12
~~3~~~~1~~2
− 178
134

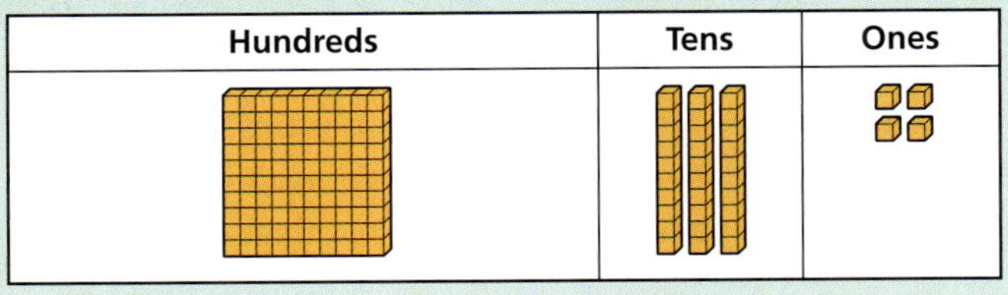

Teaching Idea **2.43**

Ask students to describe two ways to calculate 51 − 37. [e.g., Add up from 37 to 51: add 3 to get to 40 and then add 11 to get to 51, so 51 − 37 = 3 + 11 = 14; Subtract 40 and then add back: 51 − 40 = 11 and 11 + 3 = 14.]

To focus on BIWNO 4, ask: *What are the advantages of each way of subtracting?* [e.g., The first way is good because adding is sometimes easier than subtracting. The second way is good because the subtractions are easy to do mentally.]

Subtraction Algorithm 2

Because addition and subtraction undo each other (**Principle 1**), you can subtract too much in order to use a number that makes the calculation easier and then undo what you did by adding back the amount you subtracted. You can perform this algorithm mentally.

97 − 58

Subtract 2 too many:
97 − (58 + 2)
= 97 − 60
= 37

Add 2 back:
37 + 2 = 39

97 − 58 = 39

414 − 296

Subtract 4 too many:
414 − (296 + 4)
= 414 − 300
= 114

Add 4 back:
114 + 4 = 118

414 − 296 = 118

Subtraction Algorithm 3

This algorithm is based on **Principle 4** (you can subtract in parts). You subtract in parts because the parts are easy to subtract mentally.

50 − 27

Subtract 27 in parts (25 + 2):
50 − 27 = 50 − 25 − 2
= 25 − 2
= 23

50 − 27 = 23

414 − 302

Subtract 302 in parts (300 + 2):
414 − 302 = 414 − 300 − 2
= 114 − 2
= 112

414 − 302 = 112

Subtraction Algorithm 4

This algorithm is based on **Principle 5** (to subtract two numbers, you can add or subtract the same amount to or from both numbers without changing the difference). By selecting appropriate amounts to add or subtract, you can transform a calculation so you can do it mentally.	52 − 32 *Subtract 2 from each number:* 52 − 32 = (52 − 2) − (32 − 2) = 50 − 30 = 20 52 − 32 = 20	414 − 296 *Add 4 to each number:* 414 − 296 = (414 + 4) − (296 + 4) = 418 − 300 = 118 414 − 296 = 118

Teaching Idea 2.44

Ask students to solve these two subtractions:

501 − 498   536 − 211

[e.g., 501 − 498 → 498 + 2 + 1 = 501 so 501 − 498 = 3; 536 − 211 = 536 − 200 − 11 = 336 − 11 = 325]

To focus on BIWNO 4, ask: *Why might you do these two subtractions differently?* [e.g., It is easy to count up to subtract for 501 − 498 but not for 536 − 211.]

Subtraction Algorithm 5

This "making change" algorithm uses the relationship between addition and subtraction (**Principle 1**). When you make change, you count up from the price (the subtrahend) to the money paid (the minuend) to find the difference. Although you often do the counting up mentally, you must still add the "change"; some students need to jot down these amounts so they do not forget them.	57 − 18 *Think of the difference as a missing addend:* 57 − 18 = ■ → 18 + ■ = 57 *Count up from 18 to 57:* *Add to get to a multiple of 10:* 18 + <u>2</u> = 20 *Add to get to the next multiple of 10:* 20 + <u>30</u> = 50 *Add to get to the minuend:* 50 + <u>7</u> = 57 *Add up the "change":* <u>2</u> + <u>30</u> + <u>7</u> = 39 57 − 18 = 39
	327 − 158 *Think of the difference as a missing addend:* 327 − 158 = ■ → 158 + ■ = 327 *Count up from 158 to 327:* *Add to get to a multiple of 10:* 158 + <u>2</u> = 160 *Add to get to a multiple of 100:* 160 + <u>40</u> = 200 *Add to get to the minuend:* 200 + <u>127</u> = 327 *Add up the "change":* <u>2</u> + <u>40</u> + <u>127</u> = 169 327 − 158 = 169

Fractions and Decimals

A fraction is a number that describes a relationship between a part (represented by the numerator) and a whole (represented by the denominator). Although you see two numbers, you have to think of one idea, the relationship. As a result, fractions pose a challenge for many students and even for many adults.

BIG IDEAS FOR FRACTIONS AND DECIMALS

1. Fractions can represent parts of regions, parts of sets, or parts of measures. These meanings are equivalent (e.g., $\frac{1}{3}$ of a region is 1 whole divided into 3 equal parts). (BIFD 1)
2. A fraction is not meaningful without knowing what the whole is. (BIFD 2)
3. There are always two fractions involved in any single fraction situation—the part you are considering and the rest of the whole (e.g., whenever there is a $\frac{3}{5}$, there has to be a $\frac{2}{5}$). (BIFD 3)
4. You can use a fraction to describe any partitions of a whole. Decimals describe partitions of wholes to preserve the pattern of the place value system. (BIFD 4)
5. You can describe the same portion using two or more different fractions (e.g., $\frac{1}{2}$ or $\frac{2}{4}$), or using a fraction and a decimal (e.g., $\frac{1}{10}$ or 0.1). (BIFD 5)

Each teaching idea in this section of the chapter will indicate which Big Idea(s) for Fractions and Decimals (BIFD) can be emphasized.

Fractions

The first fraction students meet is usually $\frac{1}{2}$. They usually work next with other unit fractions, such as $\frac{1}{3}$ and $\frac{1}{4}$. Once students have a firm grasp of the commonly used unit fractions, they typically extend their work to other proper fractions, such as $\frac{2}{3}$, $\frac{3}{4}$, and $\frac{5}{8}$. In later grades, they meet improper fractions like $\frac{3}{2}$, and mixed numbers such as $1\frac{1}{3}$ and $2\frac{3}{4}$. Eventually, students work with fraction operations, usually in middle school or junior high school.

Teaching Idea 2.45

Ask students to draw pictures or tell about situations that they can describe using the fraction $\frac{1}{2}$.

To focus on BIFD 1 and 2, ask: *How does your picture or situation show $\frac{1}{2}$?* [e.g., The sandwich had 2 pieces that were the same but only 1 piece was left after she ate some.] *What is the whole? What are the 2 parts?* [e.g., The sandwich is the whole; the 2 pieces of the sandwich are the 2 parts.]

Meanings of Fractions

In the early grades, students normally focus on fractions of a region, fractions of a measure, specifically length (usually represented as a position on a number line), and, in some jurisdictions, fractions of a set or group. In later grades, they are introduced to the other meanings of fractions (division and ratio).

MEANING	EXAMPLE
Region (Area)	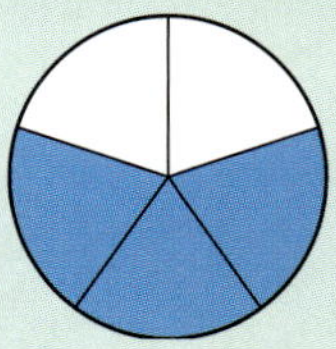 $\frac{3}{5}$ of this circular region is blue.

<table>
<tr><th>MEANING</th><th>EXAMPLE</th></tr>
<tr><td>Measure (e.g., Length)</td><td>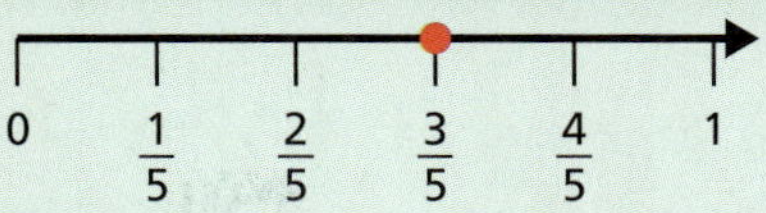

$\frac{3}{5}$ of the way along the length from 0 to 1 is marked.</td></tr>
<tr><td>Set or Group</td><td>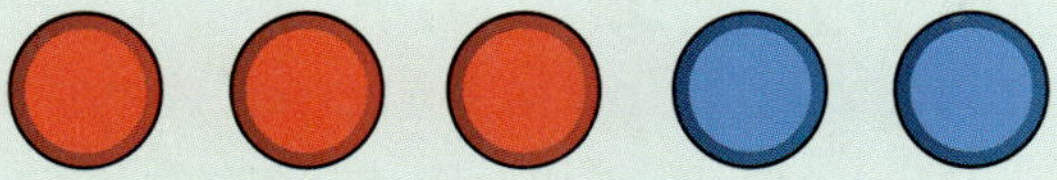
$\frac{3}{5}$ of the group of counters are red.</td></tr>
</table>

Teaching Idea 2.46

Ask groups of 5 students to describe themselves so that something is true about $\frac{3}{5}$ of the students in the group. [e.g., 3 students are standing and the other 2 are sitting, or 3 are boys and 2 are girls]

To focus on BIFD 3, ask: *What is true of $\frac{2}{5}$ of the group?* [e.g., $\frac{2}{5}$ are sitting]

How the Fraction Meanings Are Equivalent

Because a fraction can have different meanings, at some point students have to put these meanings together and see that they are equivalent. They need to see that it makes sense to apply the same fraction in each situation even though both the context and the problem situations are different. Some methods for helping students bring some of these meanings together are shown below. Here is a model that shows the relationship between a fraction of a set and a fraction of a region:

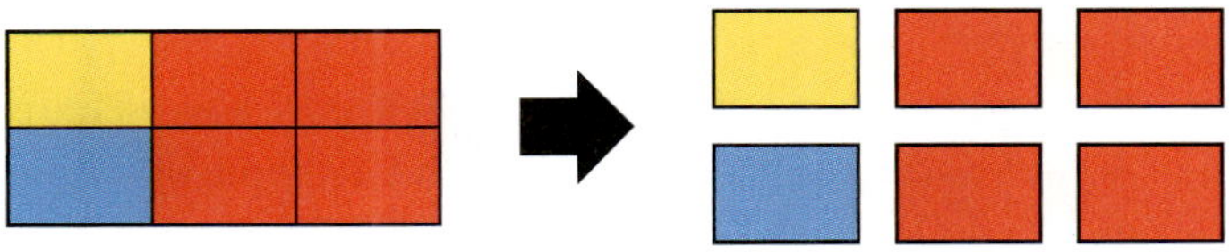

$\frac{1}{6}$ of the whole rectangular region is yellow.
$\frac{1}{6}$ of the set of smaller rectangles is yellow.

Representing Fractions

It is important for students to recognize that different fractions can represent the same amount, depending on what they consider the whole to be. For example,

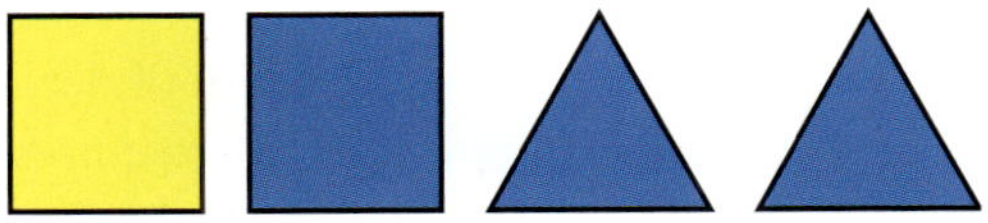

The yellow square is $\frac{1}{2}$ of the set of squares, but $\frac{1}{4}$ of the set of four shapes.

Students need to realize that whenever a fraction is represented, the fraction that represents the rest of the whole automatically appears. For example,

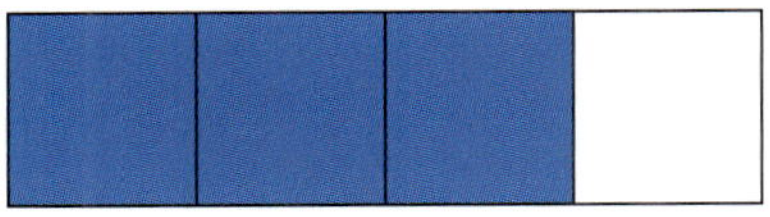

This picture shows both $\frac{3}{4}$ and $\frac{1}{4}$.

Teaching Idea 2.47

Provide pattern blocks to students and ask them to find a block that is $\frac{1}{3}$ of another block. [blue is $\frac{1}{3}$ of yellow, or green is $\frac{1}{3}$ of red]

To focus on BIFD 2, ask: *How does your block show $\frac{1}{3}$?* [3 of these blocks cover 1 of the other blocks, so 1 block covers $\frac{1}{3}$.] *How is it possible that both the blue block and the green block represent $\frac{1}{3}$?* [The block that is the whole is different for each.]

A third important notion about fractions (and decimals) is that the parts do not have to be contiguous. For example,

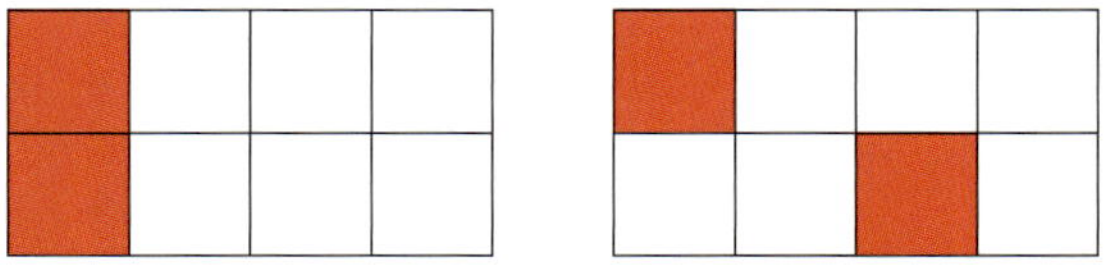

$\frac{2}{8}$ of each rectangular area is red.

Naming a Fraction in Different Ways

Although we do not introduce equivalent fractions formally at the K to 3 level, students may notice equivalence in concrete and pictorial situations. You can address equivalence in an informal way. For example, in the arrangement below, students are just as likely to notice that $\frac{1}{2}$ of the rectangle is blue (the top half) as that $\frac{2}{4}$ is blue (2 out of 4 pieces).

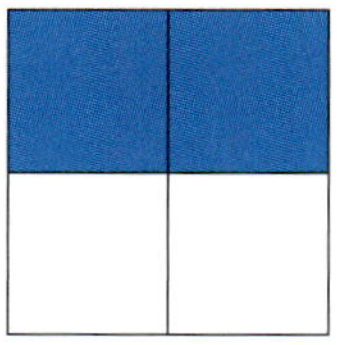

This picture shows that $\frac{1}{2} = \frac{2}{4}$ in an obvious way.

Decimals

Students will already be familiar with decimals if they have seen prices in the form of dollars and cents. However, they understand a notation such as $3.14 in terms of 3 dollars and 14 cents and not 3 dollars and 14 hundredths of a dollar, which is the underlying decimal meaning.

Although in many jurisdictions, decimals are not a formal part of the curriculum prior to Grade 4, they are introduced in some jurisdictions in Canada in Grade 3. Regardless, it is also helpful for K to 3 teachers to have a sense of decimal development to help them work with money situations in the earlier grades and also to help ensure that they use appropriate place value language with whole numbers, language that will not cause problems down the road once decimals are introduced.

Representing Decimals

It is not really possible for students to make sense of decimals unless they understand fractions. An introduction to decimals requires familiarity with the concept of fraction tenths. It is natural to use items that come in tens to begin renaming fraction tenths as decimal tenths.

Fingers are familiar and they come in tens, so they make an excellent introductory context for decimals.

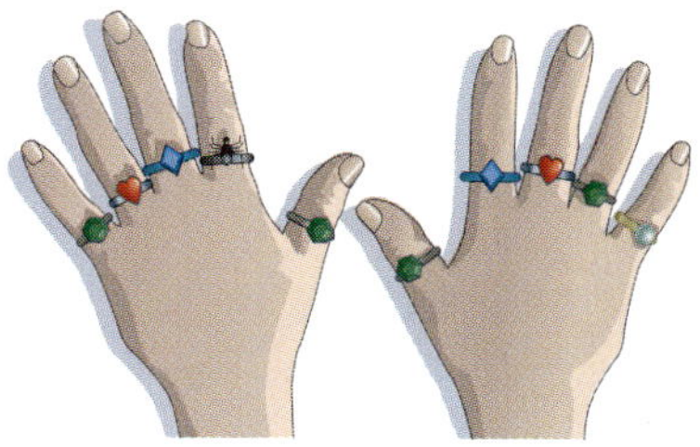

$\frac{4}{10}$, or 0.4 of the fingers have green rings.

Teaching Idea 2.48

Ask students to use coloured square tiles or linking cubes to build a rectangle that is $\frac{6}{8}$ blue, for example,

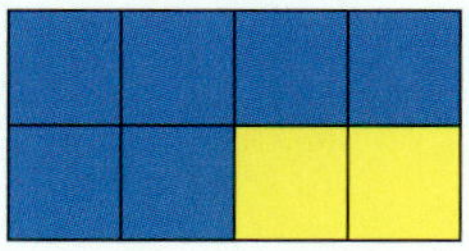

To focus on BIFD 5, ask: *How can you rearrange your tiles so it looks like $\frac{3}{4}$ of the rectangle is blue? Explain your thinking.*

[

3 of the 4 columns are now blue.]

Teaching Idea 2.49

Display a rectangle divided into 7 equal parts: 5 red and 2 blue. Then display a rectangle divided into 10 equal parts: 4 red and 6 blue.

To focus on BIFD 4 and 3, ask: *Why might someone use a fraction to describe how much of the first rectangle is red but a decimal to describe how much of the second rectangle is red?* [When something is in 10 equal parts, it's easy to name it with a decimal. If it's in seven equal parts, it's easy to name it as a fraction.] *Why can you use two decimals to describe the rectangle that has 10 parts?* [One decimal describes the red part and another decimal describes the blue part.]

However, you can show tenths using any materials that come in 10 equal sections. These might include pencils that come in packs of 10, egg cartons with 2 compartments cut off, or counters on a 10-frame.

Models for Decimal Tenths

PACKAGES OF 10 ITEMS	CONTAINERS WITH 10 EQUAL SPACES	10-FRAMES
		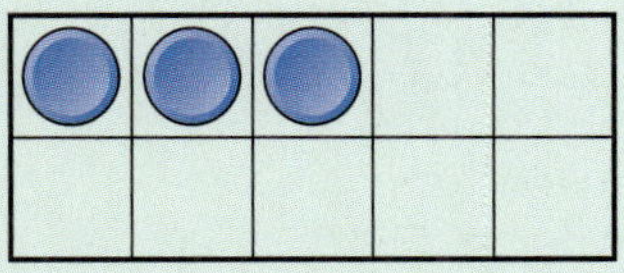
0.4 of the pencils are sharpened.	0.6 of the carton is full.	0.3 of the 10-frame is full.

Relating Decimals to Fractions

In order to discuss decimals, we need to use fraction language, for example, we say "four tenths" to read 0.4. So it makes sense to ask students first to use a fraction to describe a decimal situation. Then, because the decimal point is a convention that must be taught and modelled, you should say, "Another way to write $\frac{4}{10}$ is 0.4." You need to explain these things about decimals between 0 and 1:

- the initial 0 means there is not even 1 whole (note that it is good practice to place the 0 in front of the decimal although it is not incorrect to leave it out)
- the decimal point tells us that we are talking about parts of a whole that is made up of 10 parts
- the 4 right after the decimal point tells us we are talking about 4 of those 10 parts

Teaching Idea 2.50

Provide each student with 4, 5, or 6 counters (different students should have different numbers of counters) and ask each to draw a picture so that the counters show a decimal amount. [e.g., a sketch of a 10-frame with each counter in 1 cell]

To focus on BIFD 4, ask: *How did you make sure you could use a decimal to describe your counters?* [I used a whole that was made of 10 parts.]

Counting Using Decimals

One way to help students see how to count "one tenth, two tenths, three tenths, ..." is by showing a growing model as you say each number. Although we do not model decimals 1 or greater in K to 3, this kind of counting prepares students to recognize that we read 1.0 as "ten tenths" and that it represents 1 whole, 1.1 is "eleven tenths," and so on.

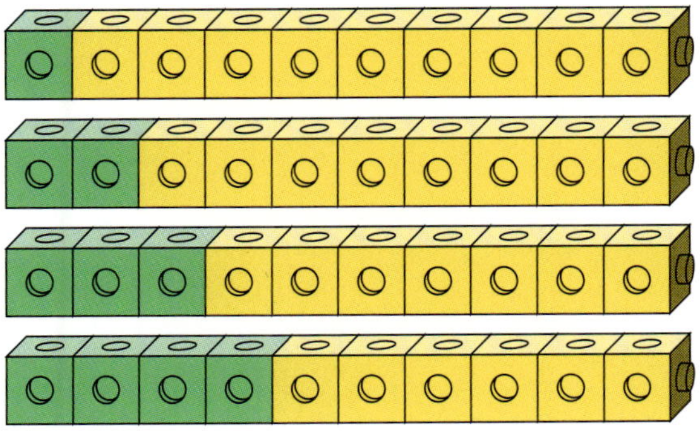

Relating to the Place Value System

You can use a place value mat or chart to help students see how decimal tenths fit into the overall place value system.

- Remind students of what they know about ones, tens, and hundreds by labelling the columns on a place value mat.
- Place 10 counters of one colour, for example, red, in the ones column and ask students what they should do [trade 10 red one counters for 1 green ten counter].

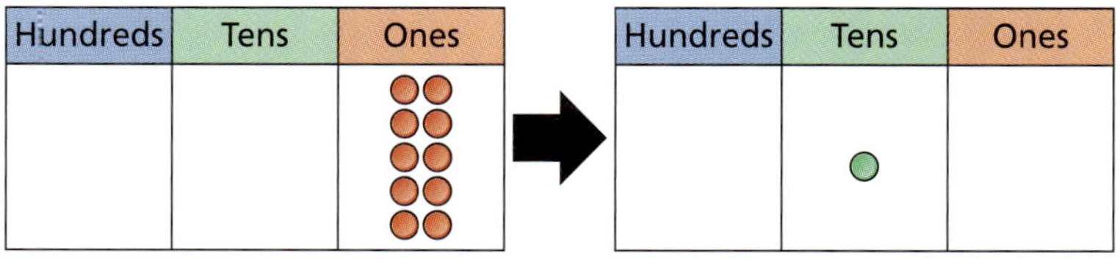

- Then introduce a new column to the right of the ones column. Call it "tenths." Place the decimal point so it straddles the ones and tenths columns (as shown below).
- Talk about how 1 is a tenth of 10, since it takes 10 ones to make a 10. In the same way, 1 tenth is a tenth of 1 because it takes 10 tenths to make 1.
- To relate this to counting by tenths, place counters of another colour, for example, yellow, one at a time in the tenths column as students count together, "1 tenth, 2 tenths, 3 tenths, …, 10 tenths." Upon reaching 10 tenths, which is a whole or 1, ask students what they should do next [trade 10 yellow tenth counters for 1 red one counter].

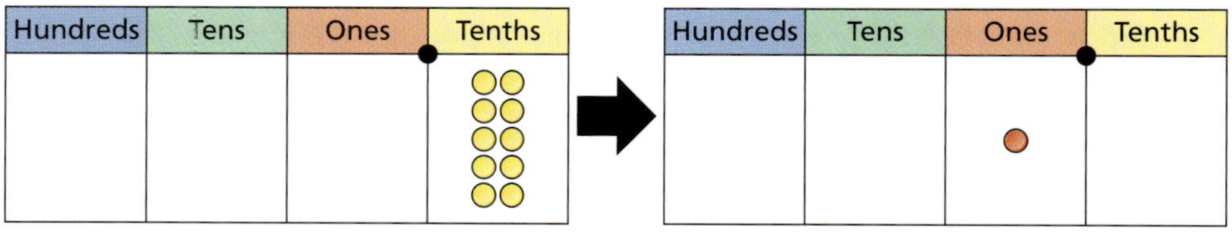

In higher grade levels, students will do much more work with fractions and decimals. A meaningful introduction with a focus on the big ideas at the K to 3 level provides a solid foundation.

Chapter 3

Geometry

Shapes and Their Properties

Many people think of geometry mostly in terms of vocabulary—do students recognize triangles, do they know what a square is, do they know what shapes are round, and so on. But, in fact, geometry is the study of spatial attributes of objects, how objects fit together, and how objects are located in space.

Geometry is the study of the spatial attributes of objects, how objects fit together, and how objects are located in space.

Attributes of shapes include things like number of sides, how pointed the corners are, whether they are round, whether they are flat (2-D), whether their parts are all the same size, and so on. It is the recognition of attributes of shapes and the implications of those attributes that helps students more effectively use shapes in their lives. Knowing that round shapes are not stable for building on is as practical to a young Kindergarten student building a structure as it is to an adult designing a real building. Representing shapes, taking them apart, and putting them together are ways to encourage students to explore more carefully the attributes of those shapes.

BIG IDEAS FOR SHAPES AND THEIR PROPERTIES

A knowledge of big ideas can help teachers choose, shape, and create tasks, and use questioning to help students make powerful connections.

1. Some attributes of shapes are quantitative, others are qualitative (e.g., the fact that a circle is round is qualitative; the fact that a triangle has three vertices is quantitative). (BISP 1)
2. Many of the properties and attributes that apply to 2-D shapes also apply to 3-D shapes. (BISP 2)
3. How a shape can be cut up (dissected) and rearranged (combined) into other shapes helps us attend to the properties of the shape (e.g., where the square corners are and whether a shape has curves or straight sides). (BISP 3)
4. Many geometric properties and attributes of shapes are related to measurement (e.g., a square is a rectangle where the width and length are equal). (BISP 4)

Each teaching idea in this section of the chapter will indicate which Big Idea(s) for Shapes and Their Properties (BISP) can be emphasized.

Teaching Idea 3.1

Sometimes it is useful to direct students when they describe the attributes of shapes, in order to get them to attend to both quantitative and qualitative attributes. Ask students to look at a variety of 2-D shapes, including several squares in different orientations and sizes and several circles in different sizes.

To focus on BISP 1, ask: *Can you use a number to describe something that is true about all of the squares?* [e.g., 4 sides] *Can you use a word to describe something that is true about all of the circles?* [e.g., round]

Attributes and Properties

As students become more familiar with the geometric attributes of various shapes, they will gradually gain an awareness of the specific attributes that define each class of shape. A specific attribute that helps define a particular class of shape is a property of that shape and applies to all shapes within that classification. For example, as students become more familiar with different types of triangles, they will eventually realize that all triangles have three sides. So if they see a shape with three sides, they will "classify" or identify it as a triangle. The focus in the early grades is on developing an awareness of the different geometric attributes; in the later grades students classify shapes by their properties.

Mathematical Language in Geometry

Students at the K to 3 level are generally not expected to use formal mathematical language to talk about shapes, but you should seize opportunities to model correct language. For example, if a student says, "The box has 8 corners," the teacher might say, "Yes, this prism does have 8 vertices, or corners."

TERMINOLOGY FOR 3-D SHAPES

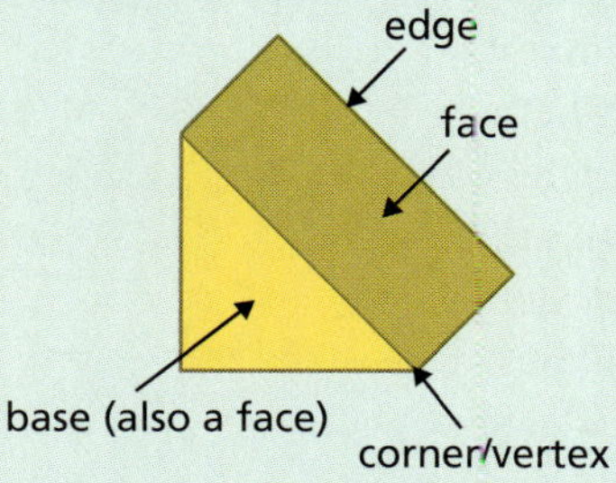

Parts of a prism or pyramid

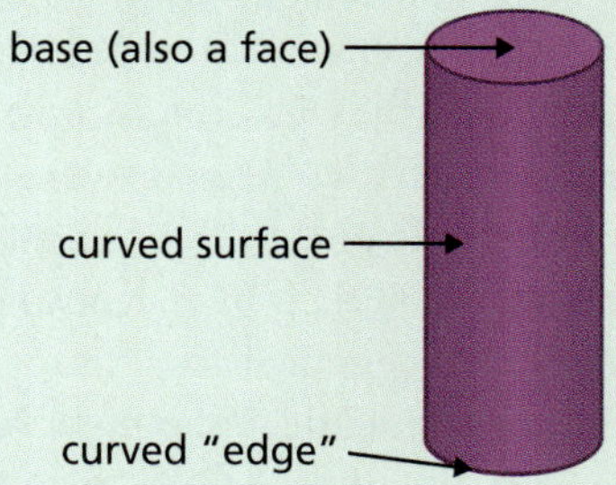

Parts of a cylinder

TERMINOLOGY FOR 2-D SHAPES

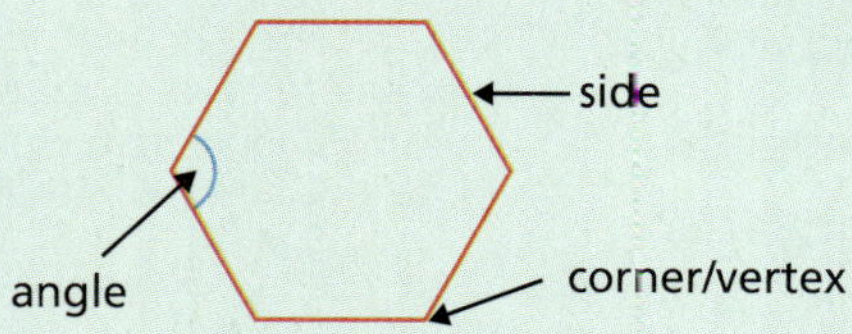

Parts of a polygon

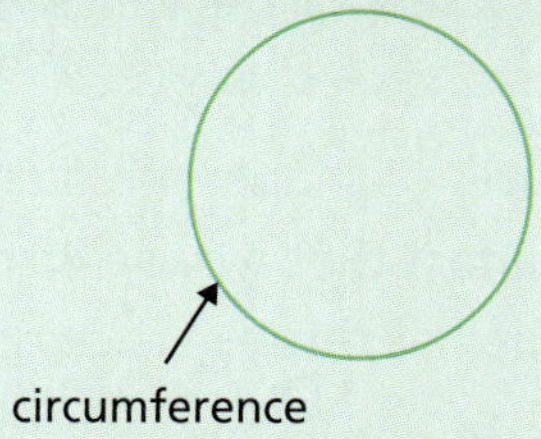

Parts of a circle

We often confuse students when we use a concrete, or 3-D representation for a 2-D shape, for example, a yellow pattern block for a hexagon.

You can choose whether to introduce the terms *3-D* and *2-D*, but it is hard to explain to a young student what the 3 and the 2 are all about. If you decide to introduce the terms 3-D and 2-D, students will get a sense that a 3-D shape has height and a 2-D shape is flat. However, we often confuse students when we use a concrete, or 3-D representation for a 2-D shape in order to allow students to manipulate the shape, for example, a yellow pattern block for a hexagon. The pattern block represents a 2-D shape, but it is actually a 3-D shape because it has three dimensions.

In some jurisdictions, teachers use the word *shape* to refer only to 2-D items. They use *figure* or *object* or *solid* to refer to 3-D items. In this resource, the word *shape* refers to both 2-D and 3-D items.

Identifying and Naming Shapes

Young students identify and name shapes on an intuitive level—they just know that a particular shape is a "ball" (sphere), a "box" (rectangle-based or rectangular prism), a square, or a triangle, although they may not recognize a shape in a non-traditional orientation, or size. For example, some students will call the yellow shape a triangle, but not the blue and green shapes.

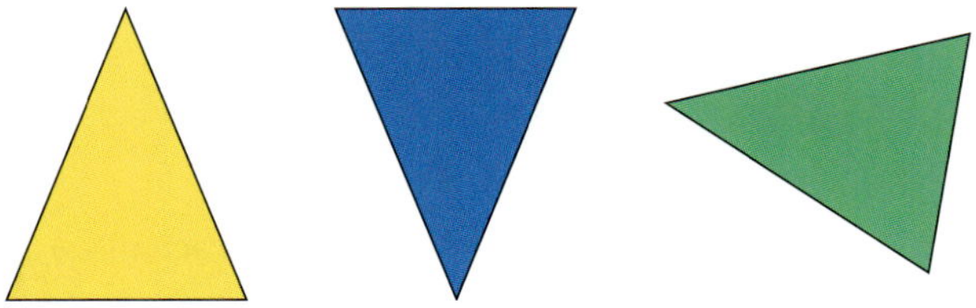

Young students may not recognize all three shapes as triangles.

It is for this reason that you should expose students to shapes in many orientations and positions. In later grades, when students refer to specific properties of shapes to identify what class of shape they are—they will know that a shape is a triangle if there are three sides and three corners, regardless of how it is turned.

Students also need to see many examples of a shape in order to identify it as a type of shape. For example, they need to see scalene, isosceles, right, and equilateral triangles so they recognize any three-sided shape as a triangle and they need to see narrow, wide, tall, and short rectangle-based prisms so they can recognize any type of rectangle-based prism as the same type of shape.

Four examples of a rectangle-based prism

Expose young students to many variations of each type of shape.

Young students should become familiar with both 3-D shapes and 2-D shapes. Generally, 3-D shapes are explored first because they are a concrete part of a child's everyday life when they see balls (spheres), boxes (prisms), cans (cylinders), and so on. And, because 2-D shapes are found on 3-D shapes, for example, squares as faces of boxes they see, it makes sense to begin with the 3-D shapes.

We often use 3-D pattern blocks to represent 2-D shapes.

Teaching Idea 3.2

Send students on a shape hunt. Create a large chart with one column for the names and pictures of the 3-D and 2-D shapes and another column for students to describe the objects they found.

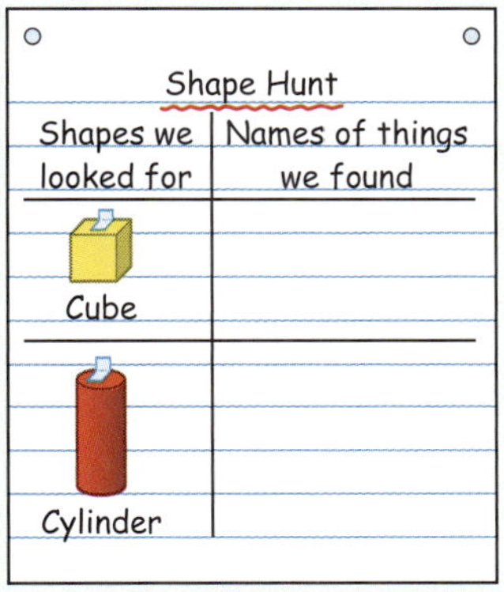

To focus on BISP 4, ask: *Why did you call one shape a square and another shape a rectangle?* [e.g., The sides of the square all looked the same, but the rectangle had some longer sides.]

Teaching Idea 3.3

Have students look at pieces of art, such as those created by Mondrian or Rothko, or modern art designs in rugs or wall hangings.

To focus on BISP 1, ask: *How are the shapes you see alike? How are they different?* [e.g., They are all triangles but some are skinny and some are not and some have square corners and some don't.]

You need to be careful to distinguish between pictures of shapes and the 3-D shapes they represent.

Students are sometimes asked to identify 3-D shapes from 2-D representations, for example, in books. You need to be careful to distinguish between the shapes that are actually on the paper and the 3-D shapes they represent. For example, in the picture below, students are not really seeing a cube, but a hexagon. When you ask what shape they see, be prepared to accept either answer.

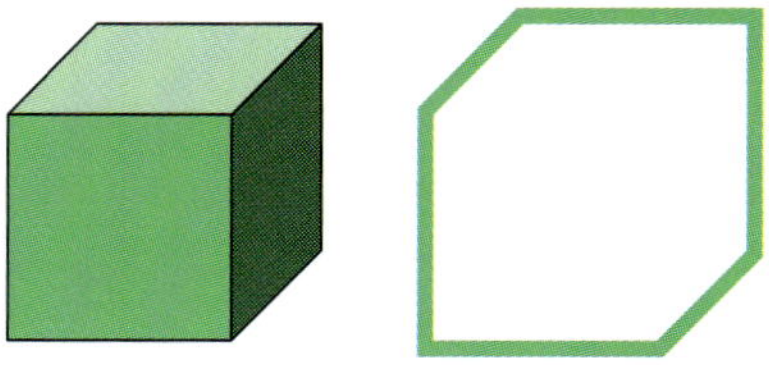

Although a picture might represent a 3-D shape, it is actually a 2-D shape.

Teaching Idea 3.4

Show students a sphere and a circle.

To focus on BISP 2, ask: *Suppose you are standing at the centre of the circle and the rest of the students in your class are standing around the outside of the circle. Which students are closest to you?* [I'd be the same distance from all of them.] *How is a sphere like the circle in that way?* [If I were at the centre, I'd be the same distance from all the parts around the outside of the sphere.] *Are there other ways spheres and circles are alike?* [e.g., They are both round.]

Many young students will name cubes as squares (or vice versa) and spheres as circles. It is likely that the similarities between the two shapes (cube and square or circle and sphere) are so overwhelming that students associate the name of the shape with one attribute of the shape. As they repeatedly hear the correct term applied to the correct shape/object and discuss how the shapes/objects differ, students will begin to make the appropriate distinctions. For example, if a child calls a cube a square, you might say, "Why does that cube remind you of a square?" or you might point to one face of the cube and say, "I agree that this part of the cube is a square."

Exploring Geometric Attributes and Properties

Young students can take part in activities to explore the attributes of shapes. As mentioned earlier, the focus at this point is on exploring and comparing shapes to become more aware of the different geometric attributes rather than on classifying shapes formally using their properties. For example, for 3-D shapes, they might consider whether or not the shape can roll or how many corners it has. For example, for 2-D shapes, they might consider whether all the corners of the shape look the same or how many sides it has.

Here are some activities that help students focus on geometric attributes:

- comparing shapes
- sorting shapes
- patterning
- representing shapes
- combining shapes
- dissecting shapes

Here are some geometric attributes that young students will observe:

3-D ATTRIBUTES	2-D ATTRIBUTES
square or triangle faces	long and short sides
number of faces or edges	number of sides or corners/vertices
identical (congruent) faces	sides the same length (congruent sides)
number of corners/vertices	pointy or square corners/vertices
round parts	round parts
more corners/vertices than faces	an even number of corners/vertices
	sides going in the same direction (parallel)

Comparing Shapes

It is useful to have students compare 2-D shapes to other 2-D shapes, compare 3-D shapes to other 3-D shapes, and compare 2-D shapes to 3-D shapes. Comparisons can draw out either similarities or differences in geometric attributes. For example, students might observe, as shown below, that some shapes are the same in some ways but different in other ways.

When comparing shapes, students are likely to notice similarities, which leads to the notion of grouping, or classifying shapes.

Comparing Shapes

COMPARING 3-D AND 2-D SHAPES

A sphere is like a circle because both are round, but a circle is flat and a sphere is not.

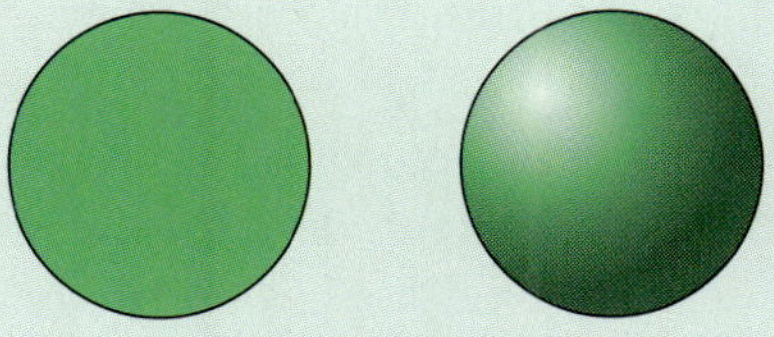

COMPARING 2-D SHAPES

A rectangle is like a square because both have the same kind of corners, but the rectangle is longer and thinner than the square.

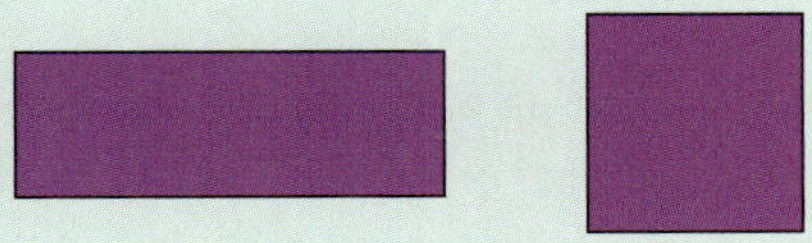

COMPARING 3-D SHAPES

A rectangle-based prism is like a cylinder because both can sit flat on a table and both have 2 identical bases, but they are different because one has round parts and the other does not.

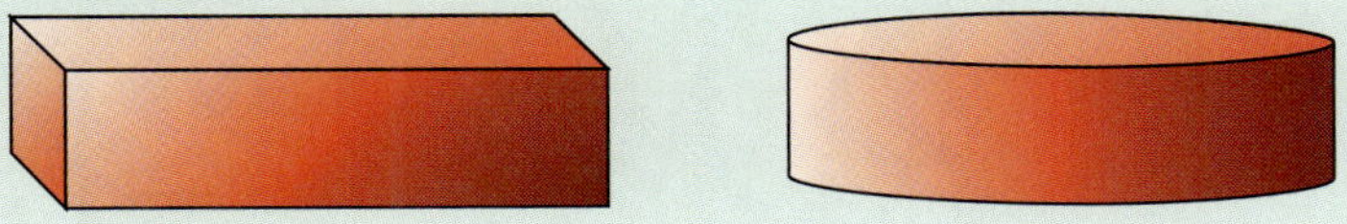

Teaching Idea 3.5

It is important for students to recognize that certain geometric attributes apply both to 2-D shapes and to 3-D shapes. Ask students to compare a triangle to a square-based pyramid.

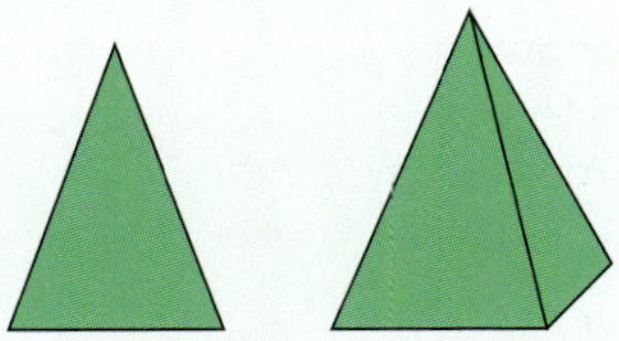

To focus on BISP 2, ask: *How are the triangle and the pyramid alike?* [e.g., They both have points.]

Besides the more obvious comparisons, another interesting way to compare shapes is in terms of the shadows they produce. You can do this using a flashlight in a darkened room or an overhead projector. For example, a cylinder can produce a circle shadow, but a prism cannot.

Students in Grades 2 and 3 might be ready to compare the shapes that make up the faces of prisms or pyramids. For example, students might trace the faces of a triangle-based prism and compare them to the faces of a square-based prism. They are likely to notice that both have some rectangle faces, but only the triangle-based prism has triangle faces. They are also likely to notice that the square-based prism has more faces.

COMPARING THE FACES OF A TRIANGLE-BASED PRISM WITH THE FACES OF A SQUARE-BASED PRISM

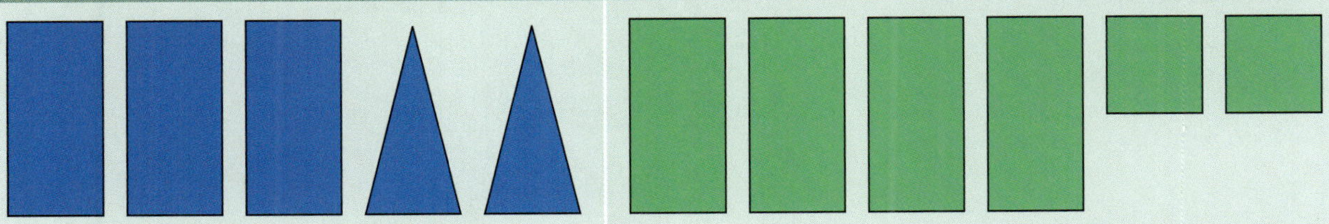

The 5 faces of a triangle-based prism

The 6 faces of a square-based prism

Teaching Idea 3.6

Provide students with a collection of 2-D shapes that includes several shapes with 2 equal sides. For example, each green shape in this collection has 2 equal side lengths:

To focus on BISP 4, ask: *How can you use measurements to help you sort the shapes?* [e.g., I can make a group of shapes that have 2 sides that are the same length.]

Sorting and Patterning

Most curriculums in Canada require elementary school students to sort and pattern with 2-D and 3-D shapes. One of the main reasons to sort shapes and create shape patterns is to get students to focus on geometric attributes that they will eventually use to sort and classify the shapes. For example, sorting a group of 2-D shapes to separate shapes with three sides from shapes with four or more sides prepares students to eventually classify shapes as triangles because they have three sides; creating an AB pattern of squares and rectangles prepares students to eventually classify squares as special rectangles with all four sides the same length. When asking students to sort or pattern with shapes, it is sometimes useful to use shapes that are all the same colour so that students use spatial attributes rather than colour as a focus when they sort and pattern.

Sorting Shapes

In the diagram below, some prisms have been sorted according to how many faces they have. By sorting in this way, students begin to discover the geometric properties of triangle-based and rectangle-based prisms, for example, all triangle-based prisms have five faces and all rectangle-based prisms have six faces.

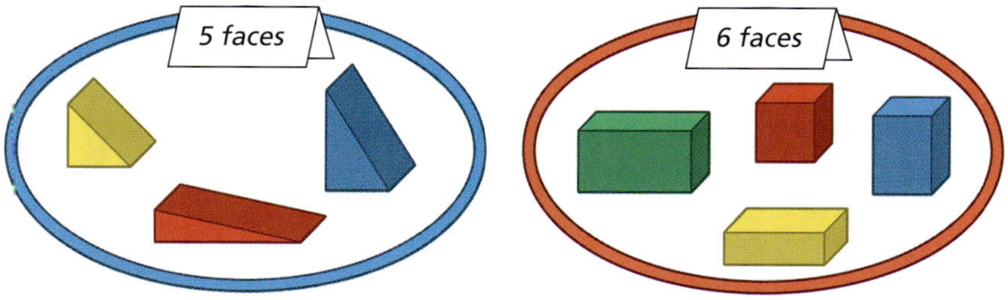

This sort shows that 5 faces are a property of triangle-based prisms and 6 faces are a property of rectangle-based prisms.

Teaching Idea 3.7

Play a game where you show 3 different rectangles and 1 triangle. Ask which shape does not belong and why. [the triangle; it has 3 sides and the others have 4 sides]

To focus on BISP 1, ask: *What is the same about all the rectangles?* [They all have 4 sides and 4 corners, and the corners are all the same.] *What is different?* [Some are almost like squares and some are long and skinny.]

You should sometimes provide students with a set of pre-sorted shapes and ask them to determine the sorting rule. For example, you might show an arrangement like the one below and ask students why the yellow shape is not inside the circle.

Why is the yellow shape not inside the sorting circle?
Which other shapes could you add to the sorting circle?

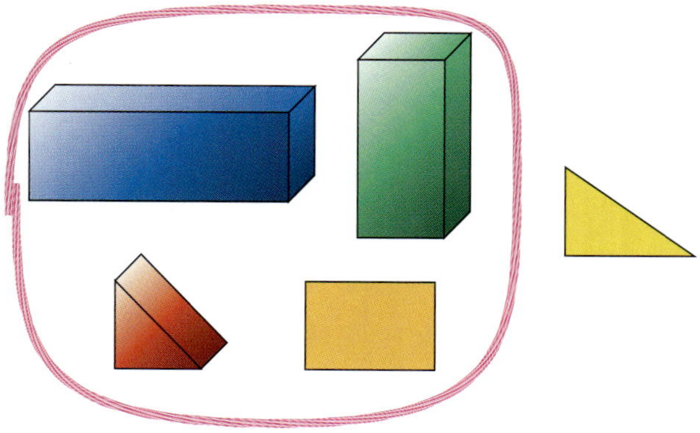

"The triangle doesn't belong because it has only three corners. I would add a yellow pattern block to the circle."

Commercial materials that are suitable for sorting and patterning include pattern blocks, attribute blocks, and 3-D solids.

Sorting and Patterning Materials

PATTERN BLOCKS	ATTRIBUTE BLOCKS	3-D SOLIDS
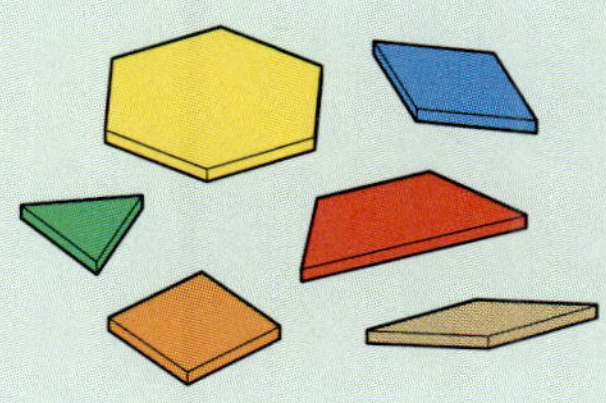		
There are 6 shapes, each a different colour.	There are 5 shapes in 2 sizes, 2 thicknesses, and 3 colours.	You can sort 3-D solids using many geometric attributes.

Patterning

Shape patterns are founded on experience with sorting and classifying according to attributes. To prepare for patterning, students can sort items using attributes such as the direction in which a shape points or properties such as number or lengths of sides (for 2-D shapes) or faces, which are flat, versus curved surfaces (for 3-D shapes). These activities help students focus on what makes one element in a pattern like or different from another. Then they can create, identify, describe, compare, and extend patterns.

Shape patterns are founded on experience with sorting and classifying according to attributes.

Students usually begin with simple repeating patterns like the one shown below, which accentuates the difference between a triangle and a quadrilateral.

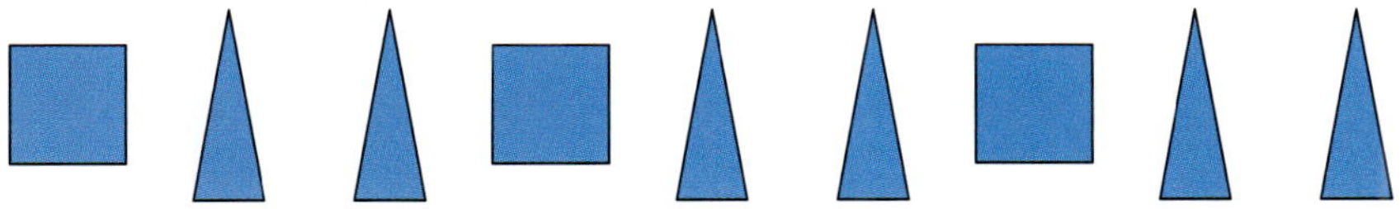

A simple repeating shape pattern with the core square-triangle-triangle (4 sides-3 sides-3 sides)

As students gain experience with shape patterns, they learn to interpret, extend, and create increasingly complex patterns. These include patterns with multiple attributes, growing/shrinking patterns, and grid patterns that change in two directions. Patterns may also involve transformations.

More Challenging Shape Patterns

A REPEATING 2-ATTRIBUTE PATTERN

In this pattern, not only do the shapes change, but the orientation of the shapes also changes.

Pattern rule 1: AAB; rectangle-rectangle-trapezoid
Pattern rule 2: AB; vertical-turned

(continued)

More Challenging Shape Patterns (continued)

A GROWING 2-DIMENSIONAL PATTERN

In this growing pattern, both the length and the width of the square increase by 1 square each time.

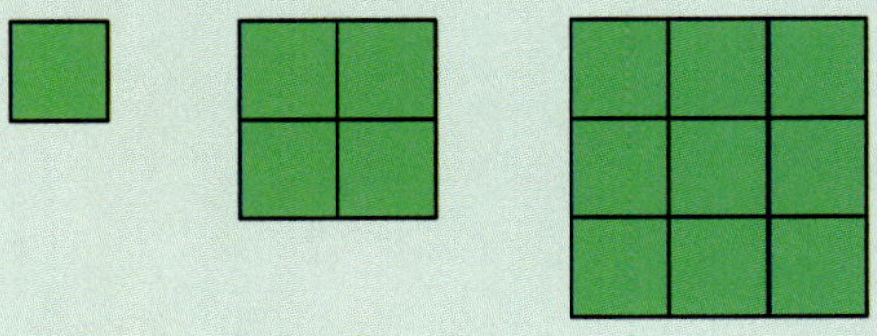

A MULTI-DIRECTIONAL PATTERN

On this grid, patterns are visible in rows, columns, and diagonals.

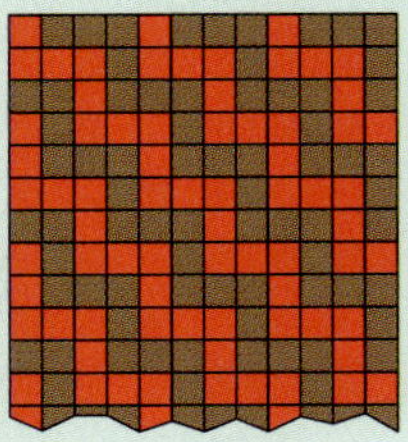

Representing 2-D and 3-D Shapes

As students represent shapes, they attend to the attributes of shapes, both 2-D and 3-D.

Modelling Shapes Concretely

One way to represent a shape is to make a concrete model, whether at a sand table, with modelling clay, or with building blocks. As they make models, students explore shape attributes or properties in a hands-on way. Many of the attributes are tactile, for example, the vertices of a pyramid are sharp points. As with any mathematics activity, students are more likely to be engaged if you present the activity in context. For example, you could have students make cookies or decorations for a special occasion.

An appropriate activity for younger students is to use modelling clay to create 2-D and 3-D shapes they can see and handle. Students can form the clay with their hands, with molds, or using cookie cutters. To create 3-D shapes, students can use geometric solids or recycled materials (cans, boxes, cardboard tubes, etc.) as models, or they can work with commercial materials such linking cubes and Polydrons. To represent 2-D shapes, students might also use geobands on a geoboard. For example, you can ask students to make large, medium, and small triangles on a geoboard, or you can ask them to create a shape with 5 corners or 5 sides.

Teaching Idea 3.8

A concrete way for students to explore the attributes and properties of 2-D shapes is to form the vertices of the shapes with their own bodies. For example, provide a group of students with a large loop of yarn. Ask 3 students each to hold the yarn and tell you what shape they have made [a triangle]. They can vary their positions along the yarn to create different triangles. Then ask a fourth student to join the group and make a rectangle.

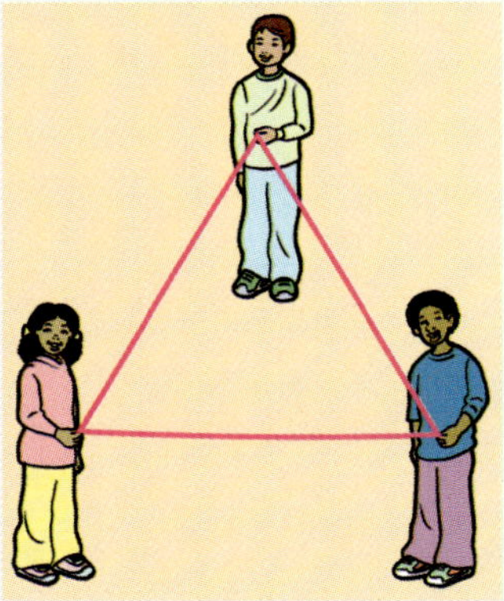

To focus on BISP 4, ask: *Why did the 4 students who formed a rectangle have to arrange themselves more carefully than the 3 who formed a triangle?* [The sides had to be the right lengths for a rectangle, but a triangle can have 3 sides of any length.]

Materials for Modelling Shapes

CLAY MODELS

POLYDRON MODELS

LINKING CUBE MODELS

GEOBOARD MODELS

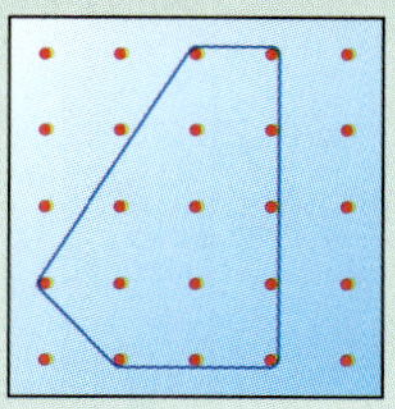

Skeleton Models

Another type of concrete model is a skeleton—a physical representation that allows students to focus on the edges and vertices of a 3-D shape, or just the sides and vertices of a 2-D shape. Materials for constructing skeletons include sticks and balls of modelling clay, Wiki sticks, or straws connected with bent segments of pipe cleaners. Toothpicks are especially useful for modelling regular shapes because they have a uniform length. Commercial construction toys such as Frameworks, Geostrips, Tinkertoys, K'nex, Zoob, Geomag, and D-stix are also suitable.

Teaching Idea 3.9

Create a skeleton of a square-based pyramid and hold it behind your back. Display several pyramid solids with different-shaped bases. Tell students that you have made a skeleton of a pyramid using 8 sticks.

To focus on BISP 1, ask: *What do you know about the faces of the pyramid?* [I know it has triangle faces because it's a pyramid, so if it has 8 sticks, the other face has to be a square.]

Materials for Creating Skeletons

GEOSTRIPS

STRAWS AND PIPE CLEANERS

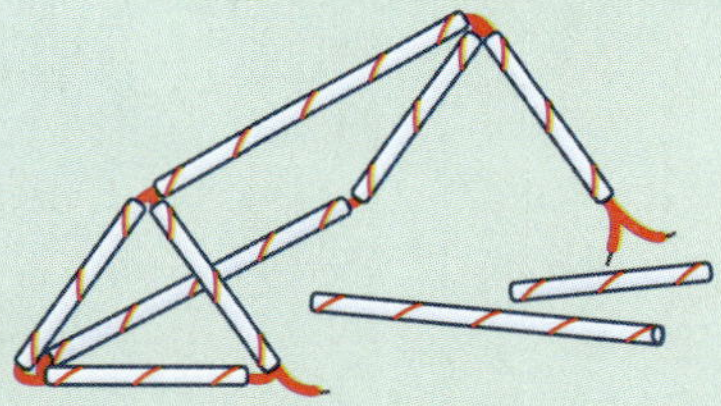

STICKS AND CLAY

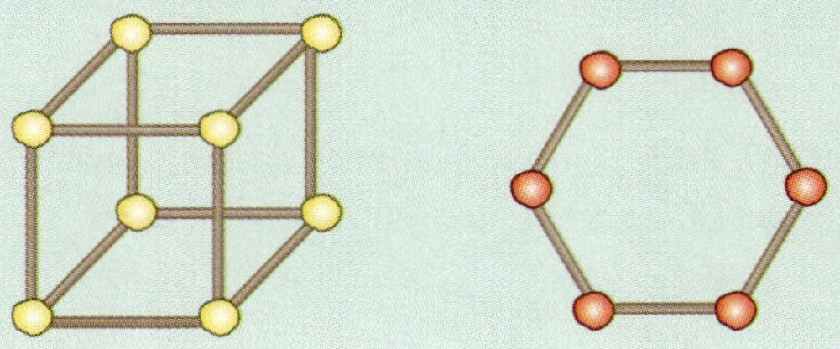

In order to construct the skeleton of a shape, many young students need to have the shape in front of them. This way, they can look at and touch the edges and vertices (focusing on the attributes of the shape) to develop a mental picture of how many there are and where they belong. Other students are comfortable working from a picture of the shape, and still others can sometimes use just the verbal descriptions.

Skeletons help students see familiar shapes in a different way. When they work with solid shapes, students tend to focus more on the faces. The process of making a skeleton, where the faces are implicit, helps students become more aware of the other components—edges and vertices. They create a mental image of the shape (visualization), which will stay with them when they no longer have concrete models to look at. For example, when asked for the number of edges on a cube, students might visualize the cube skeleton and count the edges mentally.

The process of making a skeleton helps students become more aware of the edges and vertices.

Nets allow students to focus on the faces that make up the surface of a 3-D shape.

Nets

A net is a 2-D representation of a 3-D shape that can be folded and assembled to recreate the 3-D shape. Nets allow students to focus on the faces that make up the surface of a 3-D shape—how many faces there are, their shape, and their arrangement.

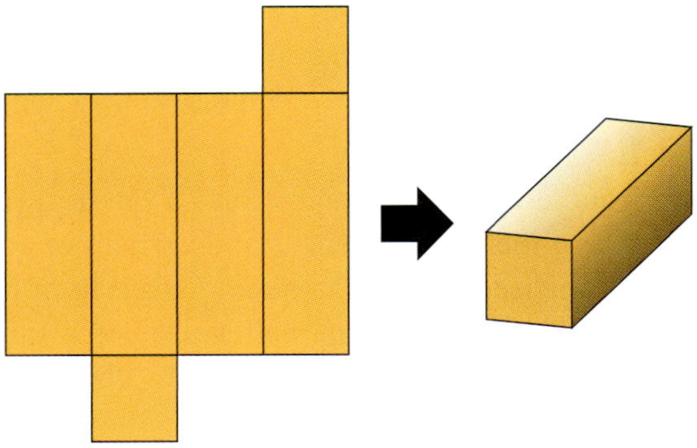

In the early grades, you might give students a net to assemble to create a simple shape, for example, a square-based prism or cube. The net might be made of plastic pieces that interlock. For example, the net below is made from Polydron pieces.

It is difficult for students to create nets on their own. However, some students might enjoy rolling a shape and tracing its faces to create a net.

Rolling and Tracing to Create a Net

STEP 1

Trace one face and mark it with a dot.

STEP 2

Roll onto another face and trace it, then mark it with a dot.

STEP 3

Roll onto another face and trace it, then mark it with a dot.

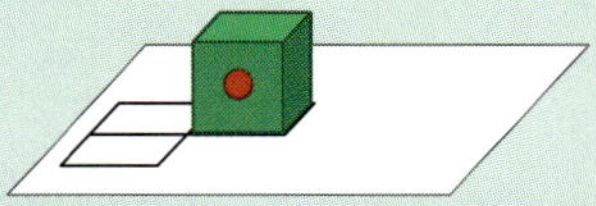

STEP 4

Roll onto another face and trace it, then mark it with a dot.

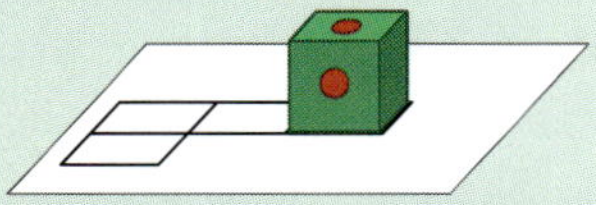

STEPS 5 AND 6

Continue rolling, tracing, and marking faces until you have traced all 6 faces. Make sure that the arrangement of squares will form a net.

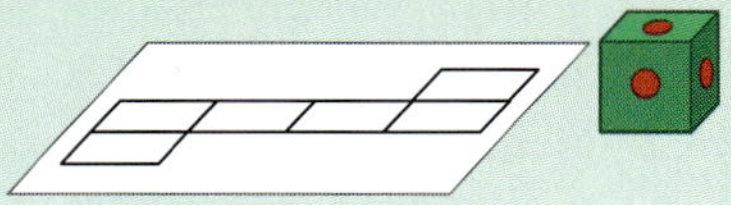

Combining and Dissecting Shapes

When students combine and dissect shapes, they attend to attributes of shapes such as side length, angle measures, and symmetry.

Combining Shapes

One way to explore the attributes of 3-D shapes is to combine them to build structures. Building structures can help students learn about how the attributes of a shape affect the way you use it. For example, a shape with faces, which are flat, can be stacked, while a shape with curved surfaces can roll or rock. Students also learn about symmetry as they build structures.

After some time building structures of their own choice, students will be ready to build a structure to match a picture or to fit a list of specifications. For example, you might challenge students to build a tall structure that is not wide, a clay model with a point, or an object made of 20 linking cubes.

Younger students also learn by making pictures with 2-D shapes (either cut-outs or stickers). As they create pictures, students focus on how the shapes they learn about in school can be combined to represent real objects in their world.

Even though a horse's body has round parts, you can use rectangles to represent the parts.

Students can also use concrete materials like pattern blocks to create pictures. For example, the illustration below shows 3 composite 2-D shapes that students can make with 4 triangle blocks.

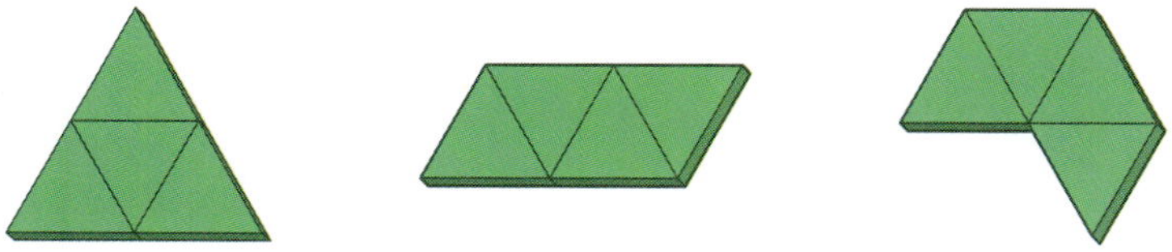

Composite shapes made from 4 pattern-block triangles

Teaching Idea 3.10

Have students model 3-D shapes by stacking pattern blocks. For example,

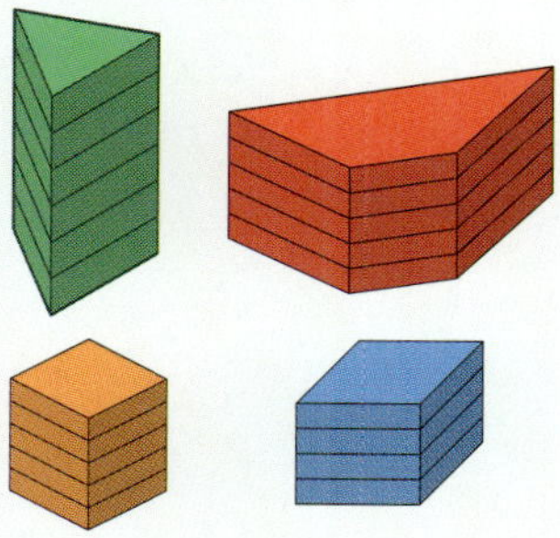

To focus on BISP 1 and BISP 3, ask: *What shapes can you make this way?* [different types of prisms] *Which shape has the most faces? The least faces?* [the stacked yellow blocks (not shown); the stacked green blocks]

Teaching Idea 3.11

Provide construction guidelines and the appropriate materials for students to build structures, for example, they can use 2 cubes, 2 prisms, 3 cylinders, and 1 cone to build a structure. Have students compare their structures with each other's and talk about the shapes that make up each bigger shape or structure.

To focus on BISP 3, hold up a 3-D shape model such as a cylinder and ask: *Can you think of this shape as being made up of smaller 3-D shapes?* [Yes; e.g., If you cut it in half, it's made of two small cylinders.]

Teaching Idea 3.12

Provide students with 4 square tiles and ask them to arrange the tiles to make a rectangle. Have students compare their rectangles.

To focus on BISP 3, ask: *Did everyone make the same rectangle? Explain.* [No; Some of us made long skinny rectangles and some made squares.] *Can you put together 3 or 5 square tiles to get a rectangle? In more than one way?* [Yes; No, only if they are all in a row.] *Do you think you can make any size rectangle with square tiles?* [e.g., Not if it's skinnier than the tiles.]

Students enjoy solving shape puzzles. One style of puzzle requires them to put pieces together to make a standard shape, such as a triangle or a square. This helps students focus on the fact that the orientation of a shape is irrelevant since they sometimes have to turn or flip pieces to finish the puzzle.

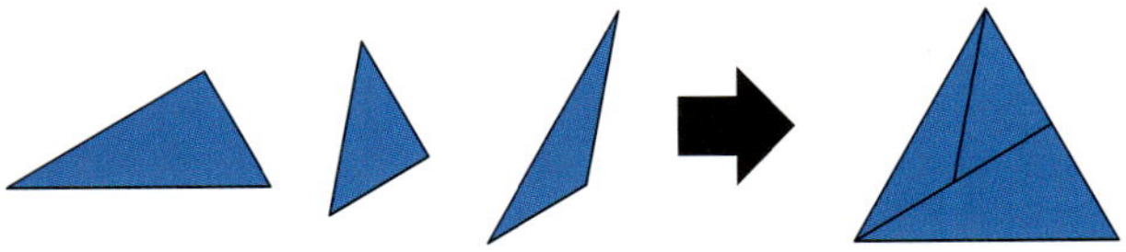

Combining shapes to complete a simple shape puzzle

Another style of puzzle has students fit shapes such as pattern blocks or attribute blocks into an outline to create a picture or to cover a design. As they work on these puzzles, students explore many geometric concepts. Students also enjoy creating puzzles like these to exchange with classmates.

As they work on puzzles like these, students explore many geometric concepts.

Pattern Block Puzzles

A VERY SIMPLE OUTLINE PUZZLE

The easiest puzzles show an outline for each block. Students place each block where it belongs on the picture or design.

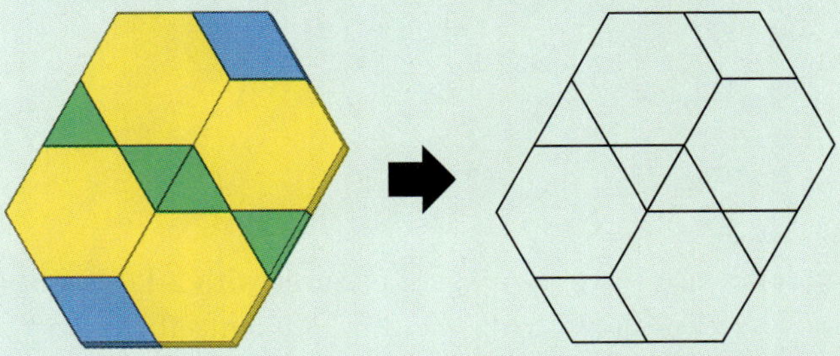

A MORE COMPLEX OUTLINE PUZZLE

More complex puzzles have outside lines for students to fill in, but the individual blocks are not outlined.

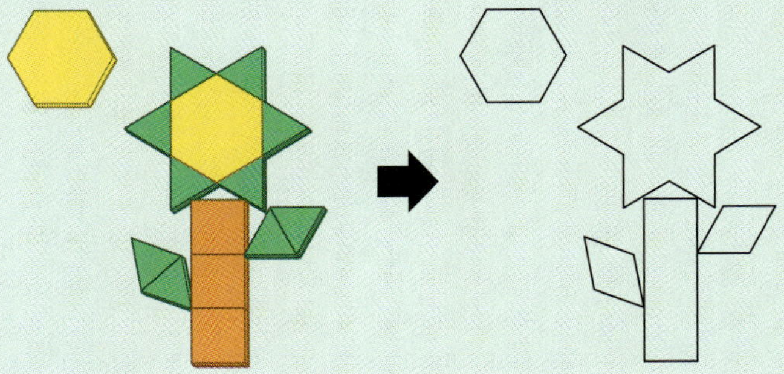

You can also use tangrams to create shape puzzles.

Tangrams

Tangram pieces (sometimes called tans) are formed by dissecting a square into seven smaller shapes as shown here. You can combine the pieces to reconstruct the original square (a challenge best reserved for older students), as well as to create many other shapes.

The seven tangram shapes

Like pattern blocks, tangrams can be used to illustrate both shape combinations and shape dissections.

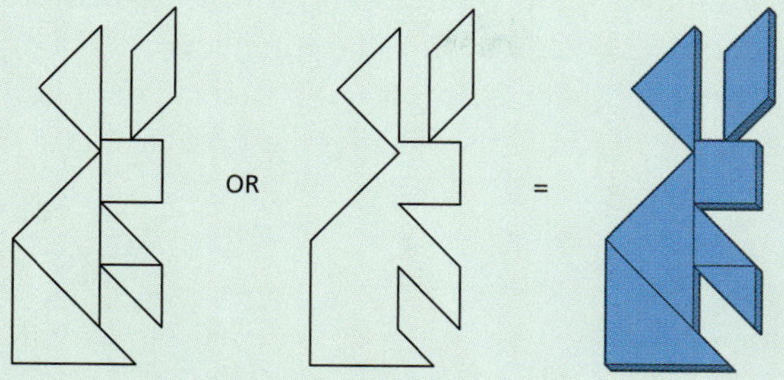

Tangram puzzles can be simple, where each piece is outlined, or more challenging, where only the outside outline is provided.

Dissecting Shapes

The idea that shapes can be dissected, or divided into parts, is fundamental to many geometry concepts students will explore in later grades.

Many interesting geometry problems revolve around dissecting a shape to create other shapes. For example, the problem below is accessible to very young students, although they will not yet be able to name all the shapes that result from the cuts.

What shapes can you make by cutting a rectangle into two parts with one straight cut?

Exploring Symmetry

In the early grades, students explore the attribute of line symmetry with respect to 2-D shapes (also called reflective or mirror symmetry). They generally decide whether a shape has symmetry by folding it to see if one half of the shape falls on top of the other half to match it completely. It if does, we say the shape is symmetrical.

Symmetrical across 2 lines of symmetry

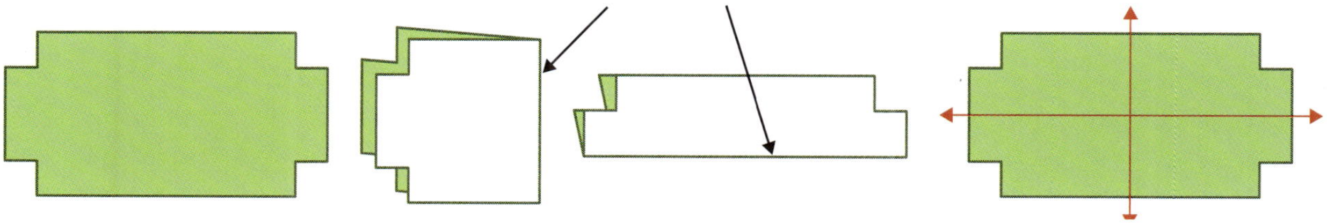

Not symmetrical

Teaching Idea 3.13

Ask students to choose an attribute or pattern block, trace it, and cut it out. Ask them to cut the shape into only triangles. For example,

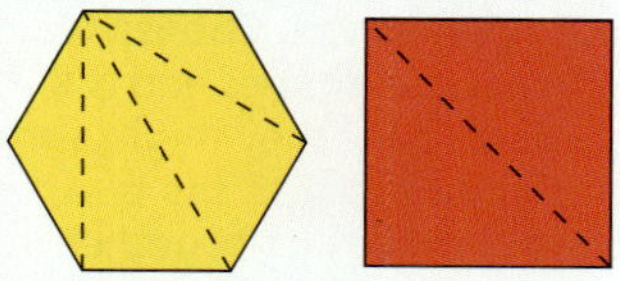

To focus on BISP 3, ask: *Can you always cut a shape into triangles? Explain your thinking.* [No; A circle can't be cut into only triangles.]

See ***page 84*** *for a discussion of the relationship between line symmetry and flips.*

Because symmetry is all around the child's world, most students expect it of a shape or object.

Because symmetry is all around the child's world, most students are not only comfortable with the concept of symmetry, but they expect it of a shape or object. For example, some young children will not accept that a shape that is not symmetrical is actually the shape in question. For example, some will argue that the blue shape below is not a triangle because it is not symmetrical, but that the red and green shapes are triangles.

Some young children might not consider the blue shape to be a triangle because it is not symmetrical.

Teaching Idea 3.14

Provide an assortment of paper 2-D shapes, including some that have 1 line of symmetry, some that have more than 1 line of symmetry, and some that are not symmetrical. Choose two symmetrical shapes and ask students how they might sort the shapes so that the two chosen shapes go together in the same sorting group.

Once students think of using symmetry, to focus on BISP 1, ask: *How do you know that all the shapes in that group have symmetry?* [I folded them in half and the halves match.] *Does it matter how many sides the shape has?* [No; e.g., Both triangles and squares can be symmetrical.] *Are some shapes more symmetrical than others? How do you know?* [Yes; e.g., I folded the square 4 ways, but I could only fold the heart 1 way.]

Young students are often comfortable completing the other half of a shape when the line of symmetry and half the shape are provided. They can also create shapes by folding a piece of paper and cutting a shape against the fold. They prefer a line of symmetry that is horizontal or vertical.

COMPLETING A SYMMETRICAL SHAPE

Folding and tracing

Using a transparent mirror

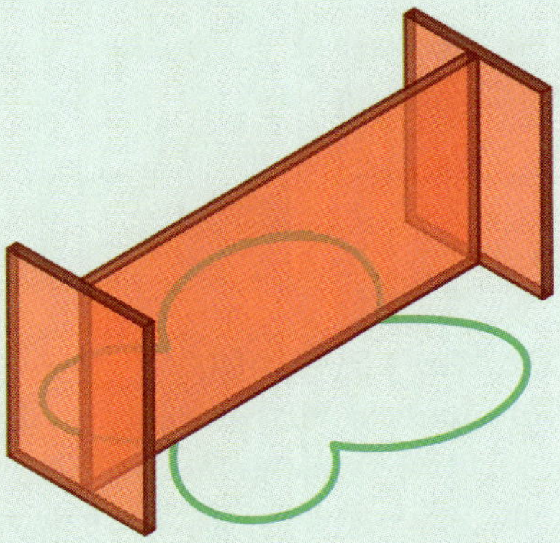

CREATING A SYMMETRICAL SHAPE

Folding and cutting

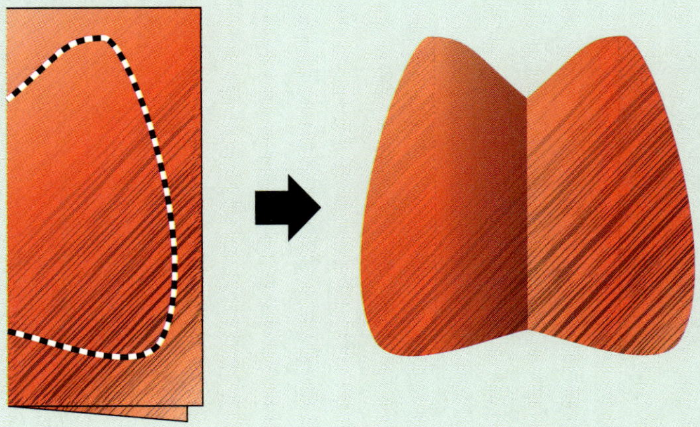

Location and Movement

Having a grasp of space involves not only focusing on attributes of individual shapes, but observing how shapes are positioned with respect to other shapes in our environment. For purposes of communication, students need positional language, such as, "the desk is *beside* the table." To predict how a particular motion affects how a shape will look or whether a shape will fit in a particular location once it is moved, students need to attend to attributes of the shape as well as to the effects of particular motions.

Having a grasp of space involves observing how shapes are positioned with respect to other shapes and attending to the effects of particular motions on shapes.

BIG IDEAS IN LOCATION AND MOVEMENT

1. Locations can be described using positional language, maps, and grids. (BILM 1)
2. Slides and flips are transformations that change the position of a shape and possibly its orientation, but they do not change its size and shape. (BILM 2)
3. Transformations are frequently observable in our everyday world. (BILM 3)

Each teaching idea in this section of the chapter will indicate which Big Idea(s) for Location and Movement (BILM) can be emphasized.

Developing Positional Vocabulary

Children's earliest spoken language might include words such as *up*, *down*, *in*, and *out*—terms that describe spatial relationships. As children grow, so does their spatial understanding and related vocabulary.

Many of the words we use to describe position are relative to the position of the speaker. For example, the same object might be *next to* one object but *across from* another object. This language is helpful in our everyday lives and is a critical part of communicating our grasp of our spatial environment.

Teaching Idea 3.15

Ask students to select an object in the classroom.

To focus on BILM 1, ask: *How can you describe where your object is so that someone can find it?* [e.g., It's next to me and also next to Andrea.] *Why would Tyler's description use different words?* [e.g., He is in a different part of the room, so the desk is not next to him.]

Positional Language in Play, Dance, and Song

Sports and imaginative play are also important opportunities for this type of development. As children play, parents and teachers can model positional vocabulary—words that describe how the location of one object relates to the location of another object.

An adult observing a child who is playing with blocks and toy farm animals might ask,

- Why did you put the cow *inside* the fence?
- Which animals are still *outside*?
- Which block can you put *on top of* this one to make your fence higher?

Many dances, songs, games, and toys for young children provide opportunities to build positional vocabulary. These activities often link well with investigations in other subject areas, such as physical education or social studies. Examples include games like Simon Says and action songs like "The Hokey Pokey."

Word Walls

Another good way to build positional vocabulary is to create a Math Word Wall to record terms as they come up in classroom activities. To help students see relationships among positional words, you can group words that belong together, such as *over* and *under*.

Our Math Word Wall for Words that Tell Where Things Are

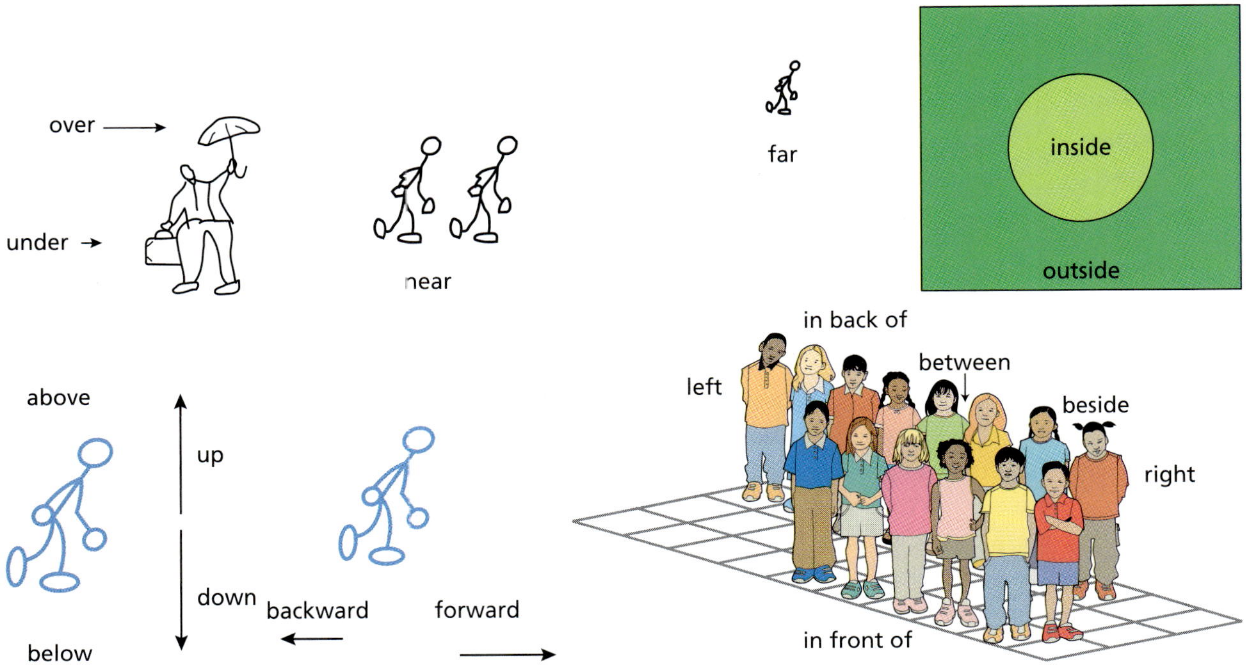

As students use and discuss these terms, they will learn that they can combine or modify terms to give a more exact idea of where an object is located. For example, younger students might discuss why it is more useful to say that the Slinky is behind the duck and to the right than to say that the Slinky is behind the duck.

Combining positional language to be more exact about location

As children mature mathematically, you can ask them to recall the relative positions of more shapes in more complex spatial arrangements. This approach to developing positional vocabulary also supports the development of visual memory as students practise recalling what they have seen once it is out of view.

Maps and Grids

Drawing and Interpreting Maps

Maps make it possible to record and describe how objects are located relative to one another. Even young students can make simple maps of their environment. As students develop better spatial sense, their maps better reflect the geometric features of objects in their surroundings and give a more accurate impression of the proportional distances between objects.

Many students enjoy figuring out paths from one location to another, for example, to get from their desk to the fire exit door or from their house to a friend's house. For example, you might ask students first to show and then to describe how to get from the bunny to the lettuce. You might ask them to show different paths they might take.

Tell how you could get the bunny to the lettuce.
Is there only one way to get there?

Teaching Idea 3.16

Encourage students to use pattern blocks to make a simple concrete map of a part of the classroom. They can use squares to represent desks, trapezoids to represent tables, and so on.

To focus on BILM 1, ask: *Why did you put the red block here instead of there?* [e.g., Because the table is in front of our desks.] *Why did you not move it farther over?* [e.g., It's not that far away from the desks.]

Working with Grids

At some point, students are ready to use a grid system to identify locations on a map or to describe how to get from one map location to another. The map below, often introduced in Grade 3 or 4, is an example of such a grid. Notice that this grid does not identify individual points, but regions or spaces. It is in later grades that students work with coordinate grids.

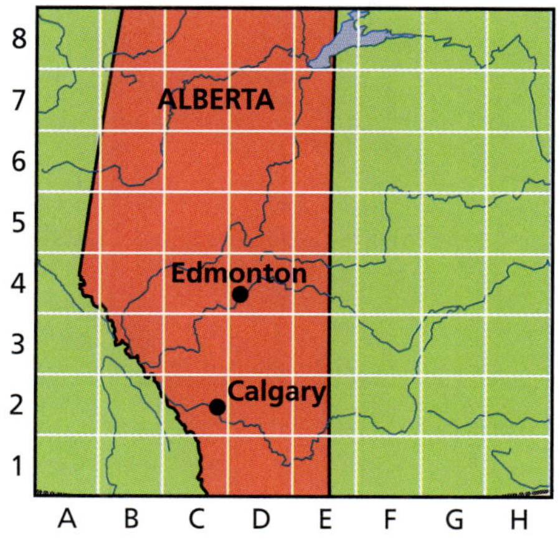

Calgary is in square C2.

Students can also create designs on grids and then describe their designs by identifying which grid squares to colour. Alternatively, you can give students grid locations and colours and ask them to show the design on a grid.

Simple 4-Quadrant Grids

Before students use grids like the grid on **page 79**, you might show them simple 4-quadrant grids like the grid below.

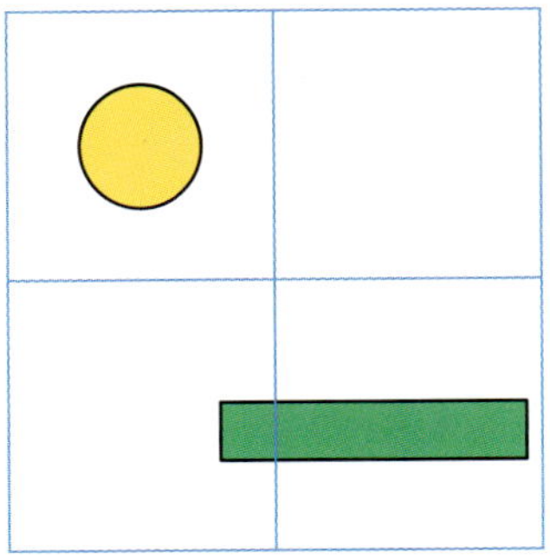

Transformations

Students work with transformations because they are the best way to describe and understand many mathematical concepts.

Sometimes shapes move and change location. Students need to think about how those shapes do and do not change when their location is changed. Often, it is slides, flips, or turns of shapes that change their locations. These motions are called geometric transformations.

A geometric transformation is a motion that affects a shape in a specified way. K to 8 students work with transformations because transformations are the best way to describe and understand many mathematical concepts, such as symmetry, and because students see many examples of transformations in everyday situations.

This rug shows examples of slides, flips, and turns. K to 3 students focus on slides and flips.

Slides (Translations) and Flips (Reflections)

K to 3 students are often exposed to slides and flips. Later, these motions are renamed as translations and reflections, and, in combination with rotations, students come to know them as the three transformations that change the location, but not the size or shape, of an object (Euclidean transformations).

SLIDE (TRANSLATION)	FLIP (REFLECTION)
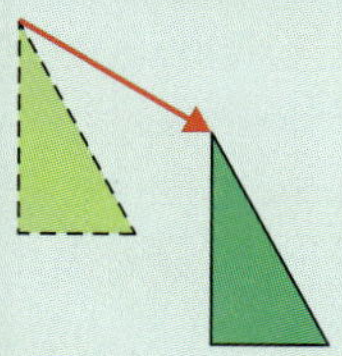 A diagonal slide ending in a position that is to the right and down	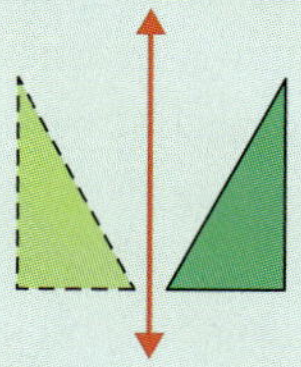 A horizontal flip across a vertical flip line

When working with transformations, it is helpful to use shapes such as scalene triangles instead of equilateral or isosceles triangles, and avoid squares and rectangles. This way, the effect of a transformation will be more obvious. For example, the motion for the transformation of the square below could have been a slide or a flip; whereas it is obvious that the motion of the crescent shape was a flip because you can tell it is facing the opposite way.

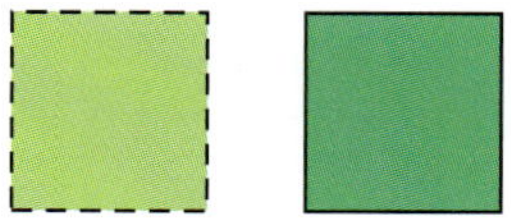

This could be a slide or a flip.

This is a flip.

Teaching Idea 3.17

Ask students to find or create a pattern by tracing a cardboard cut-out of a shape. (It is best if the shapes are somewhat irregular, i.e., not all sides or angles are equal, to make the transformations easier to recognize.) Their pattern should include flips and/or slides. Ask students to describe how the shape moved each time within the pattern.

To focus on BILM 2, ask: *How do you know that the triangle stayed the same size and shape when you slid/flipped it?* [e.g., I could put the triangles on top of each other and they would match.] *Could the size change if you slid/flipped it again? Why or why not?* [No; The whole shape moves each time, so it stays the same.]

Orientation

Informally, when we talk about a change in orientation, we mean that the shape has been turned or has changed position. This is how the term has been used in this chapter up until this point. It is also how the term is used in many elementary curriculums. Mathematically, it means something quite different and is understood by looking at examples of transformations that change and do not change a shape's orientation. One way to tell whether the orientation of a shape has changed after a motion is by comparing the order of the vertices in the original shape with the order of the vertices in its image. If the order stays the same, clockwise or counterclockwise, then the orientation has not changed. The examples below show that a slide does not change the orientation of a shape, but a flip does.

Instead of talking about a change in the orientation of a shape, elementary students will use phrases such as "Now it faces the other way." or "It's backwards now." or "It's like it would be in a mirror."

SLIDES (TRANSLATIONS) AND ORIENTATION

To read the vertices in the original image in the order ABCD, you go clockwise. The corresponding vertices in the slide image in the order A′B′C′D′ are also clockwise.

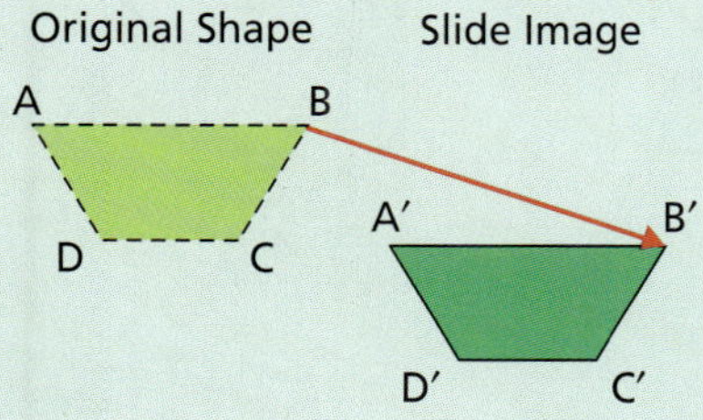

A slide image has the same orientation as the original shape.

FLIPS (REFLECTIONS) AND ORIENTATION

To read the vertices in the original image in the order ABCD, you go clockwise. But to read the corresponding vertices in the slide image in the order A′B′C′D′, you go counterclockwise.

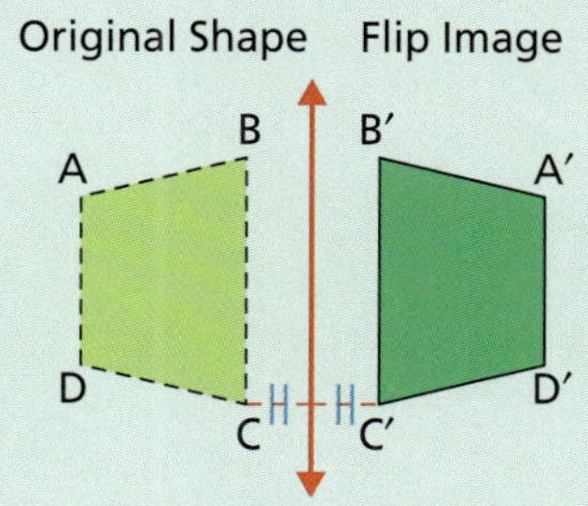

A flip image has the opposite orientation to the original shape.

Slides (Translations)

A slide (or translation) moves a shape left, right, up, down, or diagonally without changing the direction in which it faces (the orientation). This type of transformation is one of the easiest for students to recognize. For example, the pictures below show three different slides of the same triangle. Each slide is denoted by a slide arrow that links a point on the original shape to the matching point on the image. You could draw slide arrows between each pair of corresponding vertices, but one slide arrow is all that is required to show the slide.

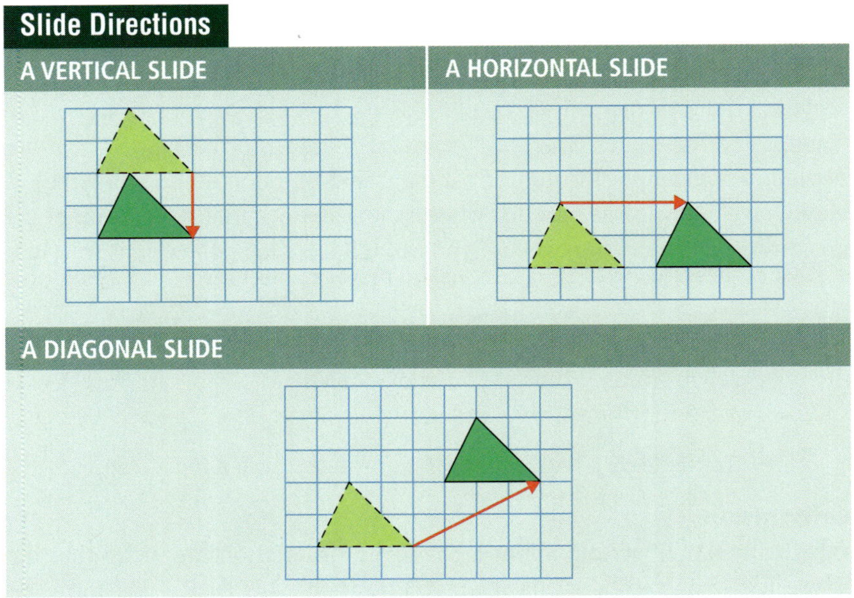

Students are often introduced to horizontal and vertical slides in the context of patterns.

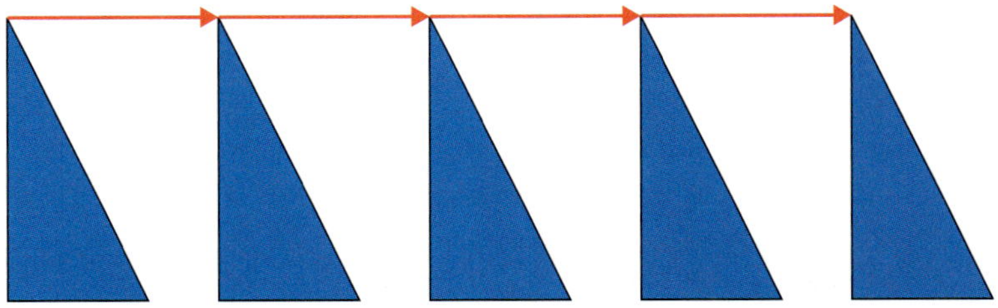

In this pattern, a triangle has been slid horizontally the same amount repeatedly.

Sometimes students meet diagonal slides in patterns.

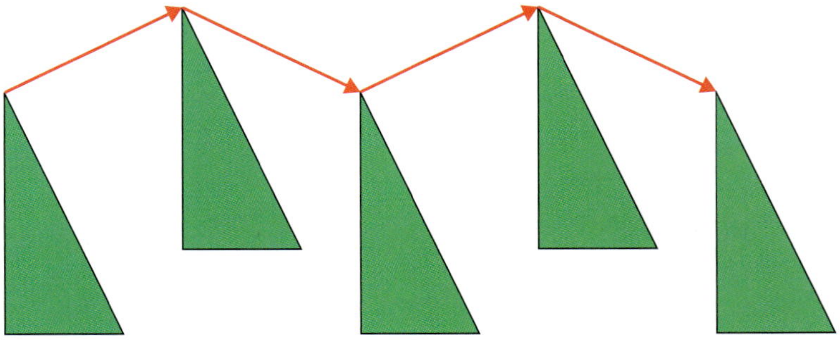

In this pattern, a triangle has been slid diagonally up and then down the same amount repeatedly.

Teaching Idea | **3.18**

Show students several samples of wallpaper, fabric, or scrapbooking paper where slides are apparent.

To focus on BILM 3, ask: *Where do you see slides in these designs? Where else do you see slides around you?* [e.g., If I look at the bulletin boards, it looks like I could slide the board on the left over and it would cover the board on the right.]

You can also use simple grids to model or describe slides. When students work with transformations, especially slides, on a simple grid, they learn to describe motions using mathematical language. For example, the light green triangle was slid to a position that is 3 spaces over and 1 space up. Even though the slide is described in two parts, the actual slide is a single motion up and to the right.

When students work with transformations on a simple grid, they learn to describe motions using mathematical language.

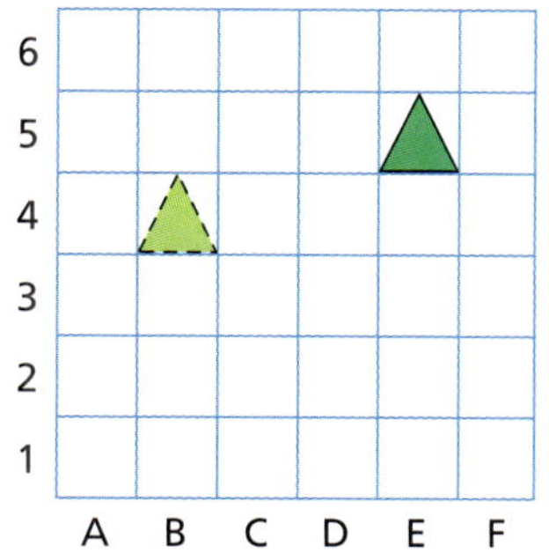

"The light green triangle was slid to a place that is 3 spaces right and 1 space up."

Initially, it is better to have students use concrete objects and actually slide them right or left, up or down, or diagonally. Students can trace initial and final positions of the shape and draw the slide arrows to have a permanent record of what happened.

Flips (Reflections)

You can think of a flip (or reflection) as the result of picking up a shape and turning it over so it faces the other way (has the opposite orientation), as shown by the front (light green) and back (dark green) of the shape below. The flip or reflection image is the mirror image of the original shape.

A flip is like turning a shape over in space.

Young students are normally more comfortable with horizontal or vertical flips, as shown below.

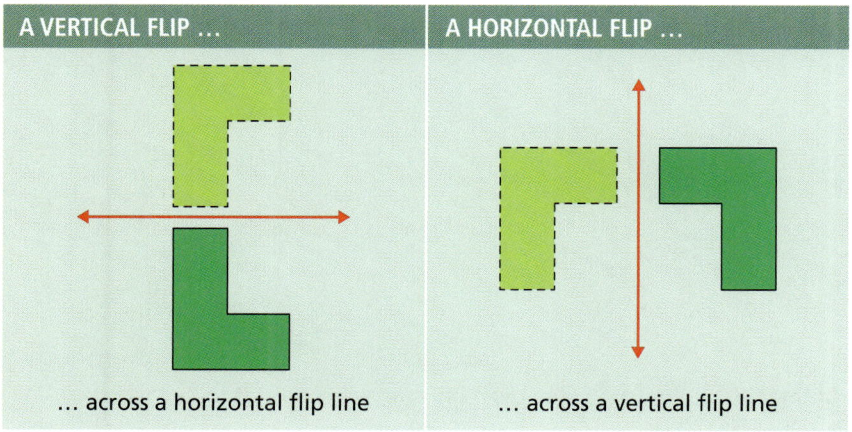

Teaching Idea 3.19

Cut out two copies of a non-symmetrical shape and flip one copy. For example,

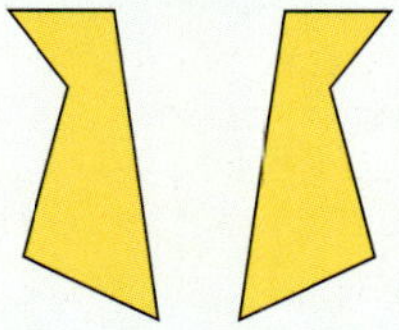

To focus on BILM 2, ask: *Did I flip the shape on the left to get the shape on the right, or did I slide it? How do you know?* [Flip; I can tell it was flipped since it's facing the other way.] *Did the size of the shape change?* [no]

Teaching Idea | **3.20**

Students enjoy seeing letters turn backwards when they are flipped. Have students write their names and use a transparent mirror to look at the reflected names.

To focus on BILM 2, ask: *When your name is flipped, can you still tell what it says? How?* [e.g., Yes; The letters are still the same, even though they face the wrong way and are in the wrong order.]

Transparent Mirrors While students generally have little difficulty flipping a shape horizontally or vertically, flips across a diagonal line can be more difficult. In this situation, a transparent mirror (or Mira) can be very helpful. Students can look through the plastic to see the flip image and then trace the image onto a piece of paper.

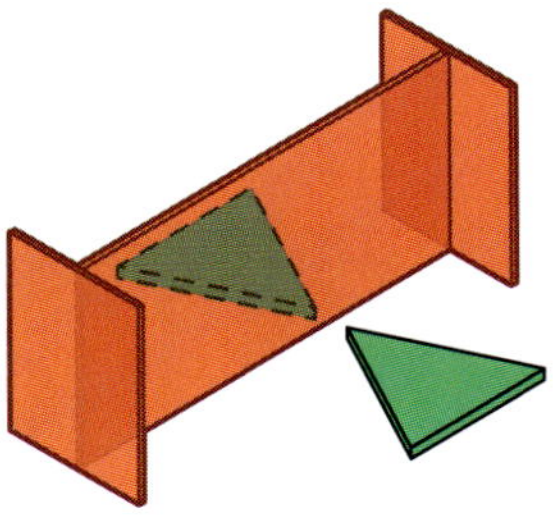

A transparent mirror is a useful tool for performing flips (reflections).

Line Symmetry and Flips

Some Grade 2 or 3 students may begin to notice that a flip creates a new symmetrical design or shape.

You can flip the shape below to make a symmetrical design.

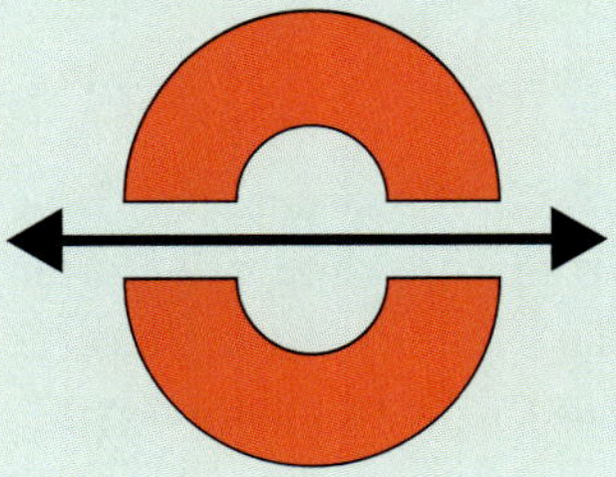

You can flip an asymmetrical right triangle to create a symmetrical triangle.

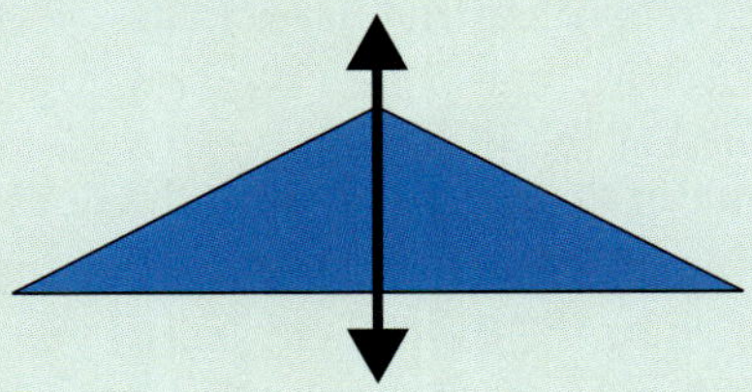

Chapter 4

Measurement

Measurement Fundamentals

Measurement is about assigning a numerical value to an attribute of an object or a situation in relation to a unit. A greater measurement implies that one object or situation has more of a particular attribute than another object.

We often describe measurements in comparative terms, for example, "Your package is twice as big as mine," where the implicit unit is the smaller package. Other times, we describe measurements using less obvious comparisons, for example, "My package has a volume of 30 cm^3," where we implicitly compare the package size to the unit 1 cm^3.

A knowledge of big ideas can help teachers choose, shape, and create tasks, and use questioning to help students make powerful connections.

BIG IDEAS FOR MEASUREMENT

1. The same object can be described using different measurements. (BIM 1)
2. Any measurement can be determined in more than one way (e.g., you can use a single unit repeatedly or use many copies of the same unit). (BIM 2)
3. There is always value in estimating a measurement, sometimes because an estimate is all you need or all that is possible, and sometimes because an estimate is a useful check on the reasonableness of a measurement. (BIM 3)
4. Familiarity with known benchmark measurements can help you estimate and calculate other measurements. (BIM 4)
5. We use units to make measurement comparisons simpler. This is only effective if the same unit is used for both objects and the unit is uniform. (BIM 5)
6. The unit chosen for a measurement affects the numerical value of the measurement; if you use a bigger unit, fewer units are required. (BIM 6)
7. The use of standard measurement units simplifies communication about the size of objects. (BIM 7)

The teaching ideas throughout this chapter will indicate which Big Idea(s) for Measurement (BIM) can be emphasized.

These seven generic big ideas underlie any type of measurement in K to 3. This chapter differs from the other chapters in this book in that it begins by describing these generic big ideas to avoid repetition in each of the sections about specific types of measurement (Length and Time, Area, and Capacity and Mass).

BIG IDEA 1 **The same object can be described using different measurements.**
Measurements can describe attributes in one dimension (e.g., length), two dimensions (e.g., area and angles), or three dimensions (e.g., mass, volume, capacity, and temperature). Time is a measurable attribute of a situation, not an object.

Any object has multiple measurement attributes. For a 2-D shape such as a rectangle, you can measure, for example, its length, its width, its area, its perimeter, or the length of its diagonal. For a 3-D shape such as a rectangle-based prism, you can measure, for example, its height, its width, its depth, the area of its base, or its capacity. For an object such as a cup of water, you could measure, for example, its mass or its temperature.

Different measurement attributes of an object are not always related, so it is possible for an object that is large in one way to be small in another way. For example, two children might "argue" about whose bowl is actually bigger if each gets a different bowl.

Teaching Idea 4.1

Present to young students a box that is very short but very wide. Let students handle the box so they can re-orient it into different positions.

To focus on BIM 1, ask: *Do you think this box is big or little? Why?* [e.g., It's big if you mean how wide it is but it's not very tall so it could be little too.]

"The yellow bowl is bigger, but the green bowl is heavier."

It is also possible for objects that are the same with respect to one measurement attribute, such as area, to be different with respect to another measurement attribute, such as width.

At the K to 3 level, the focus of the mathematics curriculum is generally on the attributes of length, area, capacity, mass, and time. Volume and angles are introduced later. Temperature is usually dealt with in science.

BIG IDEA 2 **Any measurement can be determined in more than one way.**
There are different ways to approach any measurement. For example, to measure the capacity of a bowl, you might fill the bowl with a pourable substance such as beans and then empty the contents into containers of known size. Or, you might take a measured amount of pourable material, such as a cup full of beans, and see how many times you can fill the bowl.

BIG IDEA 3 **There is always value in estimating a measurement, sometimes because an estimate is all you need or all that is possible, and sometimes because an estimate is a useful check on the reasonableness of a measurement.**

AN ESTIMATION SITUATION

There are many situations where an estimate is all that is required to solve a problem or make a decision.

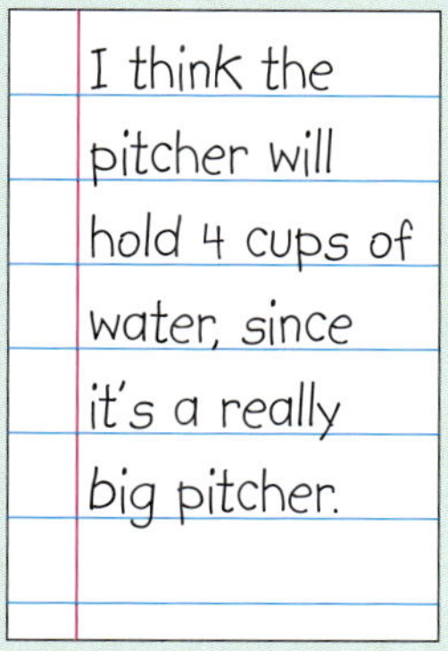

ESTIMATING TO CHECK

It is always a good idea to estimate to check a measurement. Checking an estimate often involves using known measurements (BIM 4).

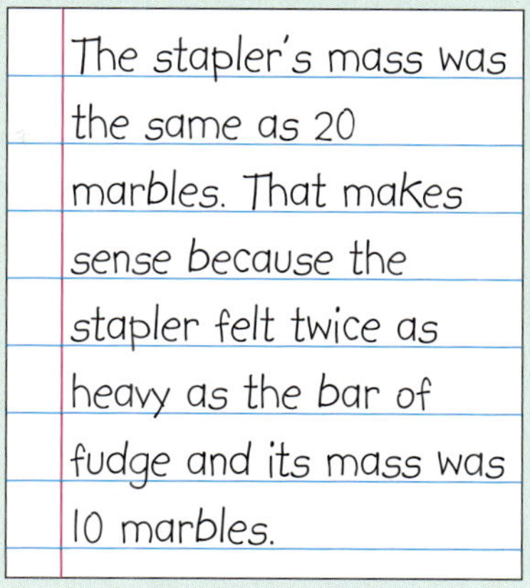

BIG IDEA 4 **Familiarity with known benchmark measurements can help you estimate and calculate other measurements.**

Each time you measure an item with a unit, you are using a benchmark. When you say that an item has a length of 10 cm, the measurement is only meaningful if you have a sense of how much 1 cm is. This big idea also describes situations when you use other known "concrete" measurements rather than metric measurements (e.g., the length of a table to estimate the length of the room).

BIG IDEA 5 **We use units to make measurement comparisons simpler. This is only effective if the same unit is used for both objects and the unit is uniform.**

One advantage of using units of measure is that they simplify comparisons. For example, without actually having the containers in front of us, we know that if one container holds 7 cups of liquid and another container holds 6 cups (of the same size) of liquid, the first container is bigger. Units allow us to compare measurements indirectly. This only works if both objects you are comparing are measured with the same unit and the unit is uniform.

Teaching Idea 4.2

Have each student choose three benchmark (or referent) lengths—short, medium, and long—that are easy to visualize, for example, the length of a thumb, the length of an arm, and the width of a classroom. Have them predict how far they could walk in 10 steps in relation to each benchmark.

Focus on BIM 4 by asking: *Is it going to be longer or shorter than your short [medium, and then long] benchmark length? Is it half as long? Twice as long?*

Comparing Using Units

USING UNIFORM UNITS

If you use identical uniform units to measure two objects, you can compare the objects using the units. For example, an object that is 4 paper clips long is twice as long as an object that is 2 paper clips long.

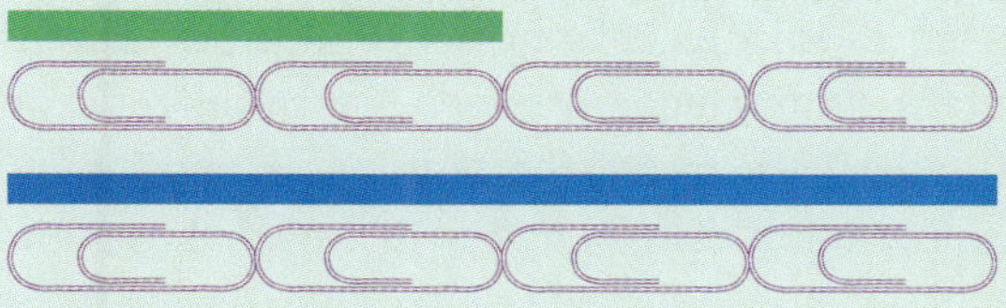

The green line is 2 units long while the blue line is 4 units long.

USING UNITS THAT ARE NOT THE SAME

If you use different units to measure two objects, you cannot compare the objects using units. For example, the measures of these two lengths have the same numerical value even though they are different lengths.

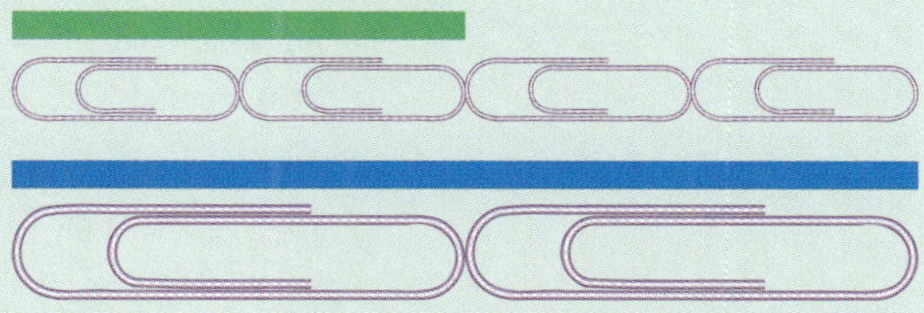

Both the green and blue lines are 2 units long, even though the blue line is twice as long as the green one.

Teaching Idea 4.3

Have students imagine that they are helping a parent prepare dessert. Their mother or dad tells them that they have to use 12 spoonfuls of water in the recipe. Show one container that holds 12 very large spoonfuls of water and another container that holds 12 very small spoonfuls of water. Tell the students that each holds 12 spoonfuls of water. Make sure students can see the two spoons but do not draw attention to them.

To focus on BIM 7, ask: *Which container of water would you use in the recipe? Explain your thinking.* [I'm not sure because both have 12 spoonfuls of water but they are different.] *How could you be sure?* [I would check the recipe again to see if it says what size spoon to use.]

BIG IDEA 6 **The unit chosen for a measurement affects the numerical value of the measurement; if you use a bigger unit, fewer units are required.**

When students know that the bigger the unit, the smaller the number, they can select appropriate units to measure. For example, you could measure the length of a room in centimetres instead of metres, but the value would be so large that it would be difficult for people to interpret the length. It is particularly important for Kindergarten or Grade 1 students, who are not yet comfortable with "big" numbers, to not use very small units that would result in such large numbers.

BIG IDEA 7 **The use of standard measurement units simplifies communication about the size of objects.**

If you know that an item has a length of 8 rods, you have no sense of how big it is unless you know how big each rod is. However, if you know that an item has a length of 24 cm, there is no ambiguity because you know how big 1 cm is.

Purposeful Measurement

It is important that measurement be purposeful whenever possible. For example, rather than simply asking a child how many pencils long a table is, you could present it as a problem to solve such as, "How could we use measurements to help us decide if the table in the library will fit at the back of the classroom?" If an authentic context seems impossible, sometimes setting up activities as games, although not as purposeful as solving real problems, engenders more engagement than context-free measurement activities. As well, asking students themselves, after they have measured something, how that measurement might be useful to them or to someone else is worthwhile.

Stages of Measurement

For most types of measurement, you can help students move through these three stages:

Three Stages of Measurement

DEFINITION/COMPARISON	NON-STANDARD UNITS	STANDARD UNITS
Students begin to define the measurement. They become aware of and apply a process for comparing items with the measurement, without using units.	Students continue to define each measurement while they measure with uniform non-standard units, such as toothpicks (for length) or small cups (for capacity).	Students learn to use measurement tools to measure with standard units, such as centimetres or litres.

Most of the work at the K to 3 level is in the definition/comparison stage and the non-standard units stage, although we usually introduce some standard units, particularly the metre, the centimetre, and sometimes the kilometre, and the kilogram and litre, and sometimes the gram and millilitre.

Do not underestimate the value of the definition/comparison stage. It is virtually impossible to define a measurement without going through this stage. To ensure students' ultimate success with measurement, it is best not to omit the non-standard units stage (except for temperature).

The Definition/Comparison Stage

In this stage, students learn what a particular type of measurement is all about by using a comparison procedure (either direct or indirect) to determine which of two items has more of that measure.

COMPARING DIRECTLY

To compare two lengths directly to see which is greater, students line them up, starting them both at the same base line, to see which extends farther.

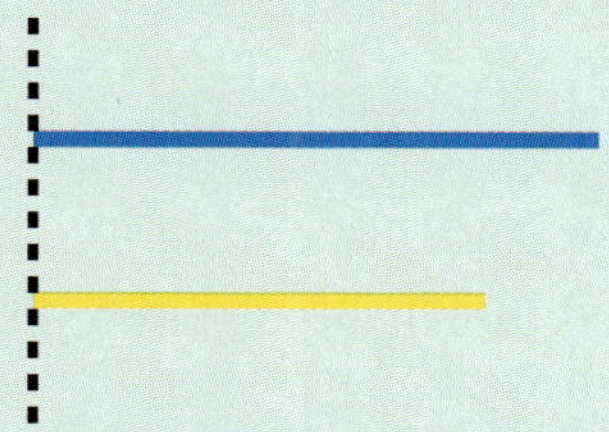

"I know the blue line is longer because the two lines start at the same place, but the blue line goes farther."

It is important for students to understand that the base line must be perpendicular to both lines. You can expose students to a situation like the one below so that they recognize this.

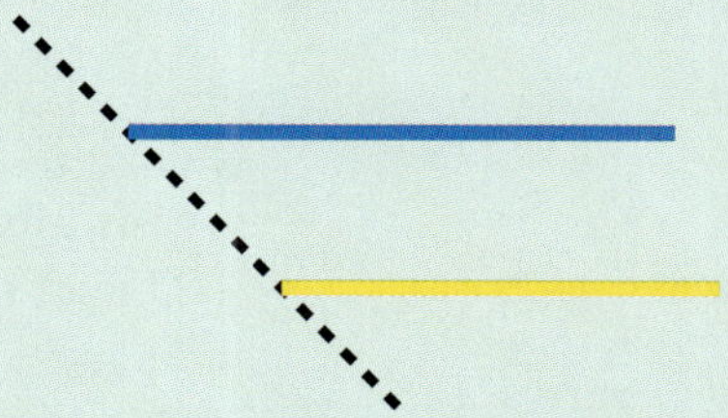

"The yellow line looks longer than the blue line but it's not, because the lines don't start at the same place."

The Non-standard Units Stage

When students use familiar objects as units for measuring, we call these non-standard units since there is no universal meaning that can be attached to them.

Using non-standard units helps reinforce what a measurement actually describes.

Using non-standard units helps reinforce what a measurement actually describes. For example, with capacity, we use a non-standard unit, like a drinking glass, rather than an abstract concept, such as the litre, to reinforce the notion that capacity is how much something holds.

You can use non-standard units to measure directly or indirectly. To use non-standard units indirectly, you measure with one non-standard unit and then "convert" to another preferred non-standard unit. For example, if you want to know the length of a book in paper clips and you know your hand span is 4 paper clips long, you can measure the length of the book in hand spans and then convert the length in hand spans to paper clips. The relationship between the two non-standard units allows you to measure indirectly.

As they work with non-standard units, students can also discover these big ideas for measurement:

Many of the Big Ideas for Measurement can be introduced and developed while working with non-standard units.

BIM 2: Any measurement can be determined in more than one way. For example, any measurement can be determined with a variety of different units.
BIM 5: We use units to make measurement comparisons simpler. This is only effective if the same unit is used for both objects and if the unit is uniform.
BIM 6: The unit chosen for a measurement affects the numerical value of the measurement; if you use a bigger unit, fewer units are required.

Another reason for using non-standard units is their practicality. Non-standard units are often easily available when standard units are not. For example, to measure mass, students can use small classroom items rather than commercial or prepared kilogram masses.

It is important that you use non-standard units that are uniform and it is helpful, particularly for young students, if they are smaller than the object being measured. Give students experiences in estimating measurements at this stage as well as in the standard units stage.

The Standard Units Stage

In the standard units stage, students use first one standard unit and eventually a variety of standard units for each type of measurement. They realize that standard units make it easier to compare and communicate about measurements. For example, if someone says that a container holds 2 glasses, it is hard to know exactly how big the container is. But if they say it holds 2 L, the size is well defined.

The Metric System of Measurement

Over 35 years ago, Canada adopted the International System of Units (called SI, for *Système International d'Unités*) as its official measurement system. Most countries around the world use this system.

One of the main benefits of the metric system is that it is consistent with the place value system.

One of the main benefits of the metric system is that it is consistent with the place value system. This makes it easier to rename units, which is of greater benefit to older students than to K to 3 students. The following chart shows how the metric prefixes relate to the base ten place value system.

Metric Prefixes

Kilo	Hecto	Deca	Unit	Deci	Centi	Milli
1000 units	100 units	10 units	1 unit	0.1 units	0.01 units	0.001 units

The most important relationships for younger students are shown below using length as the example. Note that, even though 1 cm is 0.01, or $\frac{1}{100}$ of a metre, young children will not think of the relationship in terms of hundredths. They will relate centimetres to metres as 1 m = 100 cm.

Kilometre			Metre		Centimetre	
1000 m			1 m		0.01 m	

The important metric units for K to 3 students are the metre to measure length, the kilogram to measure mass, and the litre to measure capacity. K to 3 students meet the prefix "centi" with respect to length, "kilo" with respect to mass and perhaps length, and sometimes "milli" with respect to capacity and mass, but at this stage you will probably not discuss with them the universality of these prefixes.

Length and Time

We use the term length, or linear measure, to describe measurements in one dimension. Linear measurements include height, width, length, depth, distance, and perimeter. Time, as a measurement, is about duration, or how long an event takes from beginning to end. It is grouped with length because time is a type of linear measure if you think of measuring along a timeline.

BIG IDEAS FOR LENGTH AND TIME

1. The length of an object is the number of units along a line or curve on an object. Length can be the measurement of a single measure of the object, such as the length of a rectangle, or a combined linear measure, like perimeter. (BILT 1)
2. Time is a measurement of how long something takes. Reading a clock, or telling time, is not a measurement unless you relate the time to another event. (BILT 2)

Each teaching idea in this part of the chapter will indicate which Big Idea(s) for Measurement (BIM) and which Big Idea for Length and Time (BILT) can be emphasized.

Length

Students encounter ideas about length as early as preschool because length is one-dimensional, making it the simplest and easiest to visualize of the different types of measurement, and because we often estimate and measure length in everyday life. For example, a student might want to know if a pair of pants is long enough, if a distance is too far to walk, or how the heights of family members compare.

You can introduce ideas about length using the same three stages as for other measurements: definition/comparison, non-standard units, and standard units.

Introducing Concepts of Length in Three Stages

DEFINITION/COMPARISON

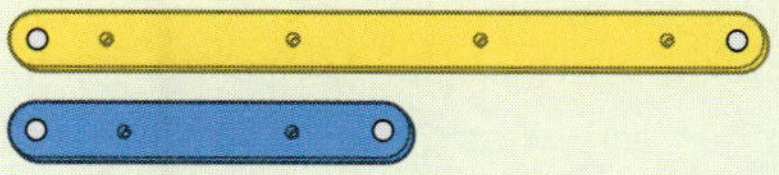

"The yellow stick is longer because it sticks out more."

NON-STANDARD UNITS

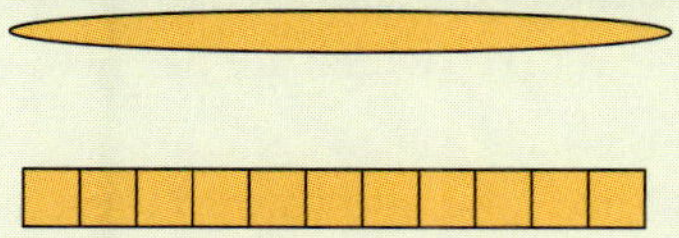

"My snake is about 11 blocks long."

STANDARD UNITS

"To make an origami frog, I have to start with a 9-cm square."

Length: The Definition/Comparison Stage

At this stage, students explore the notion that one item is longer than another if, when you line up the items to start at a common base line, one item sticks out (or up) farther than the other. In the process, students begin to develop vocabulary for comparing lengths. They use terms such as "longer than" and "shorter than." If they are comparing more than two items, they use terms such as "longest," "second-longest," "shortest," and so on.

It is important to help students recognize that the length of an object is invariant, or does not change, when you move the object. If length were dependent on position, it would be impossible to assign a value to the length of an item (see **Teaching Idea 4.4**).

Direct and Indirect Comparison

We sometimes compare lengths directly by placing the items to be compared side by side. For example, students might directly compare the lengths of snakes they have made from modelling clay to see whose snake is longest. At other times, they will need to use indirect comparison by comparing actual-size representations of objects, rather than the objects themselves. We usually use indirect comparison in situations where direct comparison would be difficult. For example, students might cut strings to represent the length of their hand-spans and compare the strings with other body lengths, such as their foot lengths or their elbow-to-wrist distances.

Students will only be comfortable with indirect measuring when they understand the transitivity of measure, that is:

If Measure A = Measure B and Measure B = Measure C,
then Measure A = Measure C.

You should give students lots of experience with direct comparison before you present indirect comparison situations.

Comparison activities, whether direct or indirect, can be part of larger investigations, such as finding out whether people with longer feet also tend to have wider hand spans.

Length: The Non-standard Units Stage

You can introduce work with non-standard units once students have realized that, rather than aligning objects directly or indirectly to compare their lengths, they can use uniform units to assign a numerical value to the length of each object. Then they can compare lengths by comparing these values.

Appropriate Non-standard Units for Length

Non-standard units are items of uniform length, such as unsharpened pencils, new crayons, new erasers, toothpicks, craft sticks (Popsicle sticks), plastic links, paper clips, base ten unit cubes, base ten rods, Cuisenaire rods, and straws. Many students use parts of their bodies, such as hand span, foot length, finger length, and arm length as non-standard units. Students should recognize that, if the units they use are not uniform, they can estimate reasonably well but they cannot be very precise in their measurement. For example, if the pencils they use are not all exactly the same length or if they measure by using the number of steps they can take (which are likely to vary from one trial to the next), precision is difficult.

For length, it is helpful to use items that are long and thin so there is no ambiguity about which attribute students are measuring. For this reason, items like recipe cards are not ideal, even though one of their dimensions could be used as a unit.

Initially, you should have students use units that are meaningful to them. For example, if they hear that something is 5 new pencils long, this gives them a good feel for its length because they already have a mental image of a new pencil's length.

Teaching Idea 4.4

Have students imagine a situation where they are very cold and want a scarf to keep them warm. Have them compare two scarves of different lengths, making sure that the shorter scarf is thicker. Compare the scarves in two different positions as shown below; each time ask students to decide which scarf is longer.

To focus on BIM 1, ask: *Why might someone say the yellow scarf is better but another person might say that the red scarf is better?* [One is longer but the other is thicker.]

To focus on the notion of invariance of length, ask: *Why might someone think that the red scarf is longer if I were to compare them this way?* [holding the scarves in the second position (so the shorter scarf extends farther to the right)] *What would you tell them?* [If they only look at the ends they might think it's longer, but it doesn't matter how you hold the scarves, the yellow one is always going to be longer.]

Teaching Idea 4.5

Use Cuisenaire rods to build a zigzag line. Ask students to use rods to build a straight line that they think is as long as the zigzag line. Provide an assortment of rods for students to select. Ask students to show how they know that the lines are the same length.

To focus on BILT 1 and BIM 5, ask: *How can you use a small rod to compare the length of the two lines?* [e.g., I can put lots of red rods on top of each line and see how long each line is in red rods.] Then ask: *How else can you compare the lengths of the lines?* [e.g., I can straighten out the zigzag line and put the two lines side-by-side to see which one sticks out farther.]

Measuring the Same Length with Different Non-standard Units

You should give students opportunities to measure the same length using different units. This will help them discover that they can describe the same item in different ways. It will also help them discover that, as the size of the unit increases, the number of units decreases. For example, the same item might be 56 paper clips long but only 8 straws long because straws are longer than paper clips.

Measuring Different Lengths with a Constant Non-standard Unit

Just as it is important for students to have opportunities to measure a length using different units, they need to measure different lengths using the same unit. This will give them measurement data that they can readily compare. It will also help build their estimation skills.

Learning to Iterate

One of the first things students learn about measuring length in non-standard units is that the units must be uniform objects used repeatedly. Initially, students measure by lining up the required number of units along the object to be measured.

Later, they learn to use a single item to iterate, that is, they move the unit from one position to another along the length of the object and count the number of times it fits along the length.

Iterating in Different Ways

MEASURING WITH MULTIPLE UNITS

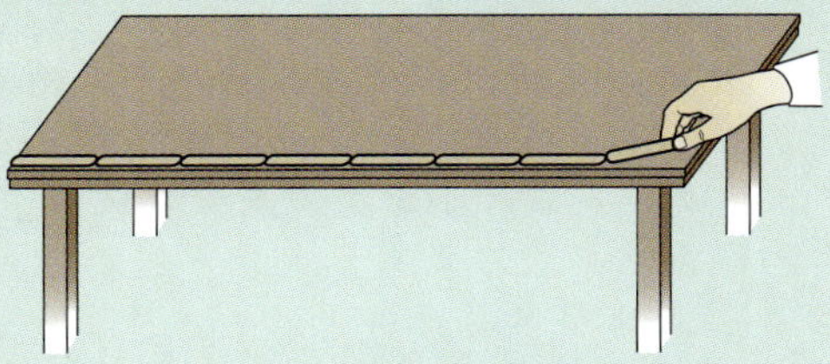

"I know the table is 8 sticks long because I used 8 sticks."

MEASURING BY ITERATING A SINGLE UNIT

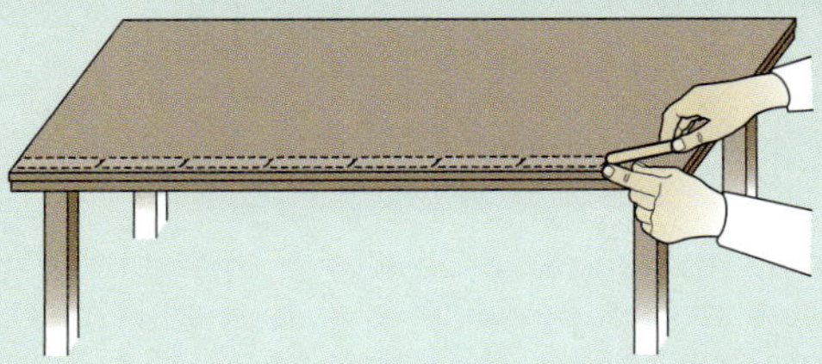

"I know the table is 8 sticks long because I used the stick 8 times."

Teaching Idea 4.6

Tell students that you are trying to figure out whether a table in your classroom is the same size as a table at home. Indicate that you have measured the length of both tables using new pencils.

To focus on BIM 6, ask: *Suppose I measure the table again using paper clips. Will I need the same number of paper clips as pencils? Why not?* [No; Paper clips are shorter, so you will need more of them.]

Now measure about half the table's length with paper clips.

To focus on BIM 4, ask: *Estimate how many paper clips long the table is. How do you know you are right?* [e.g., about 30 paper clips since it took 15 to go about halfway]

Teaching Idea 4.7

Ask students to lie down on pieces of large paper and have a partner outline their shapes. Then have each student measure his or her height using pencils.

To focus on BILT 1 and BIM 2, ask: *Do you have to measure the height all at once or can you measure it in pieces? Explain your thinking.* [e.g., I can measure my head, then the part up to my waist, and then the rest of me. Then I can add the amounts.]

Then ask students to measure how many pencils long one of their legs is and then use that to estimate how long their whole body is.

To focus on BIM 4, ask: *How did knowing how long your leg is in pencils help you estimate how long your whole body is?* [Since my legs are half my body, I just doubled the length of my legs.]

Teaching Idea 4.8

Suggest to students that you want to estimate how much longer the legs of a table are than the legs of a chair to decide if the chair is the right height for that table.

To focus on BIM 2, ask: *How can you use two different methods to figure out how much longer the table legs are?* [e.g., I can put the chair next to the table and cut a piece of string that starts where the chair leg ends and goes as high as the top of the table leg. Then I can measure the length of the string with crayons. Or, I can cut a string that is the height of the chair leg and cut another string that is the height of the table leg. Then I can measure each string in crayons and subtract the two numbers to see how much longer the table string is.]

Teaching Idea 4.9

Provide yarn to a pair of students. Ask one student, without the other watching, to shape the yarn into a curve. Challenge the other student to cut a piece of yarn that he or she estimates is the same length as the curved yarn (without putting the new piece of yarn on top of the existing piece, but only looking at it).

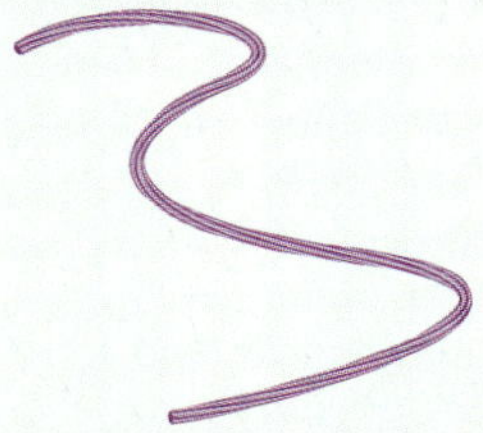

To focus on BILT 1 and BIM 4, ask: *What clues did you use to help you estimate how long the curved yarn was?* [e.g., I looked at the middle straight part of the curve and cut a piece of yarn about 3 times as long to include the two curved parts.]

Through experience, students learn that some units lend themselves to iteration more easily than others. For example, a new pencil, a paper clip, or a foot length will iterate easily, while an arm length might not.

To help students make the transition from using multiple uniform items to iteration, you can place a single long strip of modelling clay along the length to be measured so students can make end-to-end imprints of the unit until they have covered the entire distance. Students need to practise this in order to ensure that each imprint starts exactly where the last imprint left off, with no space left out and no overlap. With this form of iteration, the imprints give the impression that multiple units were used, although the length was actually measured with a single unit. Later, students will not need to do this. Instead, they will learn to use a finger as a marker to show where to start each iteration of the unit.

Another way to help with the transition from multiple units to iteration is to tell students that there are not enough units to measure the entire length and challenge students to see what strategies they can come up with if there are only two units or only one unit. If students use two units, for example, two craft sticks, there could be always a unit in place, making it easy to locate where the next unit should be placed.

Estimating Lengths in Non-standard Units

Once students have learned how to measure with non-standard units, whether using multiple units or iteration, they can estimate how many units long a particular item will be. To help students learn to estimate, you can provide situations in which an item has been partially measured. The students' task is to estimate how many units it will take to measure the full length. Eventually, students will adopt a similar strategy themselves, using partial measurements to help them make reasonable estimates.

Another opportunity for estimation is to have students use a measurement for one object to estimate the length of another object. For example, a student who knows that a doorway is 16 erasers wide can use this information to estimate the widths, in erasers, of other objects that are wider or narrower.

Students can estimate the length of Object A by measuring Object B and then using that measurement and the relationship between Objects A and B to estimate the length of Object A.

Non-standard Unit Rulers

To help students make the transition to using a standard ruler, you can have them make their own non-standard rulers by connecting uniform units, such as plastic links, linking cubes, or paper clips. These rulers help students learn that they can use the one tool to measure without having to iterate a unit. If students choose units such as foot tracings, as shown below, they will discover that lengths can be compared only if the same tool is used to measure each because of the difference between foot lengths from student to student. However, if a more "standard" unit is used, such as large plastic links, then everyone's ruler will be the same.

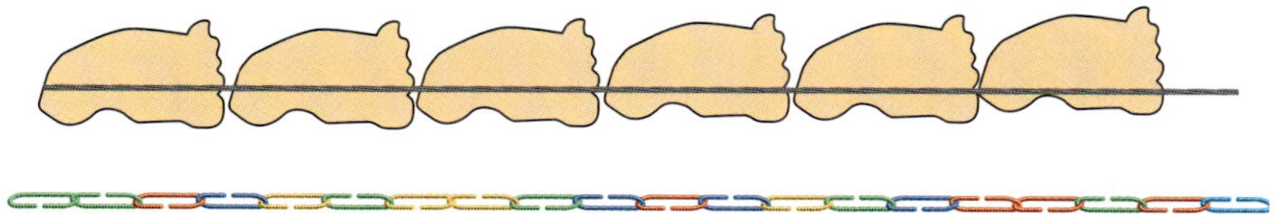

Two non-standard unit rulers

Length: The Standard Units Stage

Students can start using standard units when they realize that, rather than measuring lengths in non-standard units, which can mean different things to different people, they can instead use standard units that everyone will understand.

Introducing Standard Units of Length

It is best to introduce standard units in a context that helps students see their value and efficiency. For example, you might read a story such as *How Big Is a Foot?* by Rolf Myller (1991), which describes the confusion that ensues when a king uses his own foot to measure the dimensions of a bed he has ordered for the queen.

Educators have differing perspectives about which metric unit, centimetres or metres, to introduce first. Some believe that the metre should come first, not only because it is a base metric unit, but also because measuring in metres results in fewer errors due to problems with small motor coordination. On the other hand, it may be more convenient for the teacher if all students in a class can work on measurement activities at the same time, and this is usually only possible if students measure smaller items using the centimetre unit.

Centimetre cubes (or small base ten cubes), including base ten rods with a length of 10 cm, and metre-long sticks (unmarked initially) are useful tools for students as they begin to work with standard units of length.

Teaching Idea 4.10

Tell students that the principal has asked you to measure the length of a bookshelf in your room to see if it will fit in the storeroom. Remind students that sometimes when you want to measure something, you have to use items that are available, such as Cuisenaire rods. Ask students to use dark green Cuisenaire rods to estimate and then measure the length of a bookshelf. Then have them estimate how many light green rods long it is and then how many red rods long it is. Allow students to compare the lengths of the different rods directly.

1 dark green = 2 light green

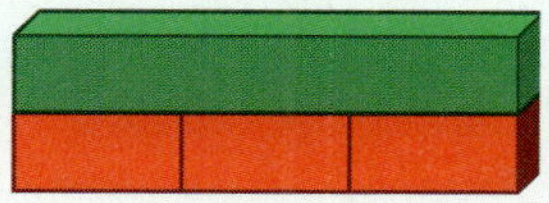

1 dark green = 3 red

To focus on BIM 4 and 6, ask: *If you know a length is 10 light green rods long, will it be more or fewer than 10 dark green rods long? Will it be more or fewer than 10 red rods long? How do you know?* [fewer dark green rods because the dark green rod is longer so I don't need as many; more red rods because the red rod is shorter so I need more of them]

STANDARD LENGTH UNITS

Introducing the Centimetre

Measuring with centimetre cubes provides a bridge from non-standard units to standard units. These act like non-standard units because students can physically line up cubes along the distance to be measured, or perhaps even link them, but they can report the measurement in the standard unit, centimetres.

Whether you introduce metres or centimetres first, you should show students what it looks like to line up 100 centimetre cubes along a metre stick.

Introducing the Metre

Many students are already familiar with the metre stick and will quickly develop a personal sense of the length of a metre. You may find that this personal sense develops more readily if students make their own tools for measuring 1 m. These tools might include lengths of string or strips of bristol board. Students can use their tools to find objects in their environment that are about a metre long, less than a metre long, and longer than a metre. The tools let them focus on the metre as a unit, without being distracted by the cm or mm markings that are usually found on metre sticks.

(continued)

STANDARD LENGTH UNITS (continued)

Introducing Centimetre Rulers

It is important to begin with a ruler that will not confuse students. Primary rulers with alternating coloured stripes work well. Students can also use rulers that show only numbered centimetres, but not millimetres. Some primary rulers have both (as shown below).

Lining up centimetre cubes along the ruler helps students see that the stripes or numbers correspond to the number of centimetre cubes that can fit along the same distance.

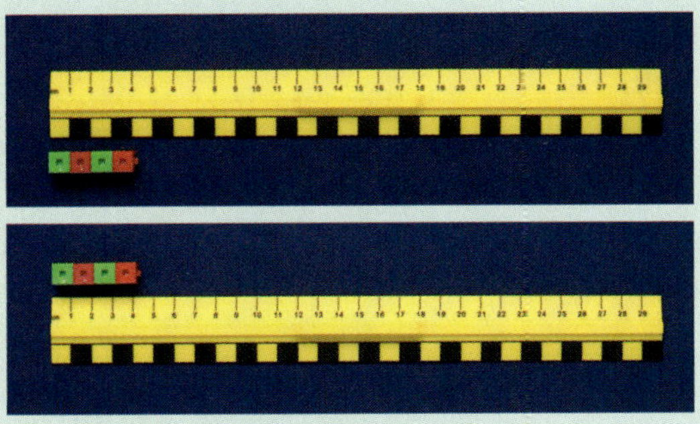

A length of 4 cm is the distance between 0 cm and 4 cm on the ruler. It is not just about the 4 cm mark on the ruler where the length ends.

Tools for Measuring in Metres

To measure lengths in metres, students can use multiple metre sticks (or measuring tapes), or they can iterate the metre stick (or tape). Measuring tapes are particularly handy for measuring lengths that are not straight lines. Trundle wheels are good for measuring longer distances in metres. A trundle wheel has a wheel circumference of 1 m, usually marked off in centimetres. As the wheel turns, it makes a click for each metre travelled. Because metre sticks, measuring tapes, and trundle wheels usually have cm markings, they can also be used to measure somewhat more precisely, for example, the room is 10 m and 17 cm long.

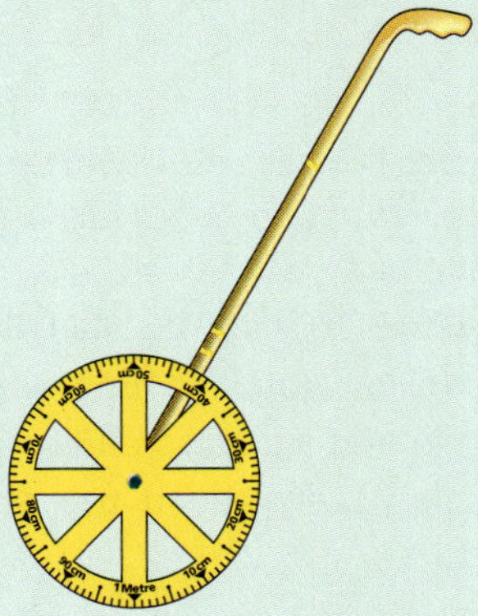

A trundle wheel

Introducing the Kilometre

Measuring with a trundle wheel is a good way to introduce the concept of the kilometre. Too often students are introduced to kilometres in an abstract way—the distance you can walk in 12 minutes or the distance between the school and the park. Students gain a more concrete and personal understanding of the size of a kilometre if they use a trundle wheel and go for a 1-km walk, counting out 1000 turns of the wheel. Even if they miscount a little, they still get a pretty good sense of the length of 1 km. Another option is to measure a 100-m running track with a 10-m piece of rope and then walk the track 10 times.

Teaching Idea 4.11

Invite two students to the front of the class. Give one a small paper clip and the other a large paper clip. Ask each student to draw a blue line that is 10 paper clips long. Then invite two more students to the front. Give each a ruler or centimetre cubes. Ask each student to draw a red line that is 10 cm long.

To focus on BIM 7, ask: *Why are the blue lines not the same length even though they are both 10 paper clips long?* [One clip was longer than the other, so 10 large clips are longer than 10 small paper clips.] *Why are the red lines the same length?* [A centimetre is always the same length, so 10 cm is always the same length.]

When students begin using rulers with numbers, you must teach them to line up the 0 mark with one end of the object to be measured. This can be difficult if the 0 mark is implicit or if the 0 mark is not right at the end of the ruler. Once students are comfortable using a ruler, you can create situations where

- the length of an object falls between two centimetre markings and they have to estimate, for example, "it is a bit more than 5 cm long" or "it is between 5 cm and 6 cm long"
- the length is greater than 30 cm and the ruler is used more than once
- the distance to be measured is not a straight line, so students must measure indirectly with string, use a measuring tape, or add the different lengths of piece
- the ruler is "broken" and does not start at 0. Students must determine the difference between two lengths by measuring from one number on the ruler to another, for example, from 3 cm to 8 cm, to deduce the length is $8 - 3 = 5$ cm

Referents for Estimating Standard Units

Referents are familiar objects of particular lengths that students can use to help them estimate. For example, a centimetre is about the width of a child's little finger and a metre is about the height from the floor to a doorknob.

Eventually, students internalize many of these referents so they can call upon the visual images when they estimate. Many educators recommend that you help students find referents that are personally meaningful, since these are easier for them to remember and eventually internalize or visualize.

Measuring Perimeter

Although people often think measuring perimeter is distinct from linear measurement, it is really only a variation in which students measure a linear distance that is not a straight line. Students can explore the perimeter of an object first using non-standard units and then using standard units when they are ready.

Situations that might make sense for young students to explore perimeter include

- deciding if one picture would take more ribbon to go around it than another picture (see **Teaching Idea 4.13**)
- deciding if it would take more chalk to draw a border around one play space than another
- deciding if it would take more small pebbles to build a border around one garden than another

MEASURING INDIRECTLY

To start, teach students to measure the perimeter of a shape by fitting a string around the shape, cutting it to that length, and then measuring the length of the string. At first, students may find it more comfortable to refer to the perimeter as the "distance around" the shape, rather than using the more formal term, perimeter.

Measuring perimeter indirectly with a string helps students see that perimeter is a type of length measurement—it is the length around an object.

MEASURING AND ADDING SIDE LENGTHS

Later, have students measure each side of a shape individually and add the side lengths to calculate the perimeter.

Teaching Idea 4.12

Encourage students to develop several personal benchmarks for centimetres and metres. They can then use their benchmarks to estimate lengths of various items.

To focus on BIM 4, ask: *How can you use your benchmark for 1 m to estimate the width of the room?* [I can imagine how many of those lengths I could line up to get across the room.]

To focus on BIM 3, ask: *Why might it be helpful to estimate the width of the room?* [e.g., to figure out how many desks will fit across]

Teaching Idea 4.13

Provide students with two pictures, one a 15 cm-by-5 cm rectangle and the other a 10 cm square. Tell students that you are planning to frame each with ribbon. Ask students to predict whether you need more, less, or the same ribbon for both pictures.

To focus on BILT 1 and BIM 1, remind students that you call the distance around a shape the perimeter and then ask: *Is the perimeter a kind of length measurement? How do you know?* [Yes; It's four lengths added together.] *Can two shapes that look different have the same perimeter? How do you know?* [Yes; Both pictures had the same perimeter but they were different rectangles.]

Time

Telling time has little to do with actually measuring time, except when reading a clock at the beginning and end of an event to figure out its duration.

Just as students learn to use tools like rulers to measure length, they need to learn to use tools like clocks and calendars to measure time. Note that reading a clock or telling time has little to do with actually measuring time, except in circumstances where the student must read a clock at the beginning and end of an event to figure out its duration.

Before they begin to measure time, students can sequence events in the order in which they occur. Although there is no time measurement per se, students must think about time in terms of what must happen first and then next. When students sequence pictures of events by arranging them in a line, they create a very simple timeline.

Once students are familiar with standard units for time, they can create standard unit timelines to help them calculate and estimate elapsed time. Although a timeline is linear, students can cut it out and form it into a circle.

Teaching Idea 4.14

Ask students to describe some things that take a long time and other things that take a short time.

To focus on BILT 2, ask: *How can you convince me that something takes a short time?* [e.g., It takes a short time if it takes less time than it takes me to sing the alphabet song; I start singing the alphabet song in my head when I start the activity to see if the activity ends before the song does.]

You can turn a timeline into a time circle to reflect the cyclical nature of time.

You can introduce ideas about time using the same three stages as for other measurement concepts—definition/comparison, non-standard units, and standard units.

Introducing Concepts of Time in Three Stages

DEFINITION/COMPARISON

When students sequence familiar events they create a rudimentary timeline.

"After I wake up, I eat breakfast. Then I walk to school."

Students compare durations by starting two events simultaneously and comparing the end times.

"I know it took my sister longer to get ready and walk to school because we woke up at the same time but she got to school after me."

NON-STANDARD UNITS

Students use tools such as hourglasses or pendulums to measure time in non-standard units.

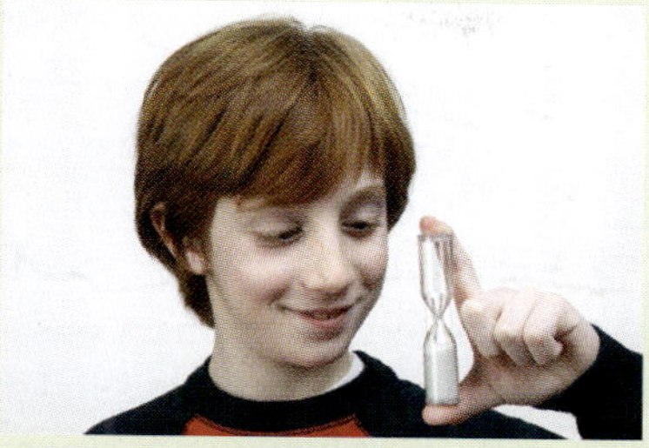

"My mom timed me while I got dressed. She turned the egg timer over 4 times from when I started until I was done."

STANDARD UNITS

Students measure time in minutes and hours by measuring elapsed time directly using standard units or by calculating elapsed time.

"It took me 1 hour and 13 minutes to get ready and walk to school."

Woke up Arrived at school

"It took me from 7:30 to 8:43 to get ready and walk to school—that's 1 hour and 13 minutes."

Time: The Definition/Comparison Stage

When students sequence events by ordering a set of pictures, they are creating a rudimentary timeline.

At the definition/comparison stage, we normally focus on which event took more time, but teachers often have students sequence events to create a rudimentary timeline. This will help the students later on when they must think about what happens first when they measure time. Younger students often enjoy drawing or organizing a sequence of pictures to tell a story. The story might be about what happens before and after school, or after school and before supper. You might also have students tell oral stories about what happened first, then next, then next, and so on.

In terms of measuring time at this stage, students compare the durations of two events to see which takes longer. Without units, the only way to compare durations is to start both events simultaneously and wait to see which event keeps going when the other is over. Many students will see the parallel to measuring length; a longer event "extends farther" than a shorter event.

In the definition/comparison stage, students build vocabulary to talk about time. They might tell what happened *first, last, next, before*, or *after*, or they might identify one event as *longer, shorter, faster*, or *slower* than another.

Time: The Non-standard Units Stage

Although in the early grades we introduce students to standard units such as hours and minutes, it is still helpful for them to work with non-standard units.

Although in the early grades we introduce students to standard units such as hours and minutes because these are used in everyday life, it is still helpful for them to work with non-standard units to measure duration before they begin using standard units to measure time. A broad experience with measuring in different ways helps students build a better understanding of what it really means to measure time.

There are many non-standard tools that students can use to measure duration, or elapsed time, including sand timers of various sizes, pendulums, and metronomes (online metronomes are available). For example, a student could count the number of pendulum swings to measure how long it takes someone to walk down the hall and back.

An advantage of many of the tools used for measuring time in non-standard units is that you can adjust them to measure time in a wide variety of unit sizes. For example, you can set a traditional metronome for different speeds. You can adjust a pendulum by using a longer or shorter string, and you can make sand clocks with varying amounts of sand and different dimensions. As students make these adjustments, they build an understanding of the factors that influence how the size of the unit affects the resulting measurement.

Teaching Idea **4.15**

Play a piece of music that has both a fast beat and a slow beat. Talk to students about how musicians choose to mix up speeds in their music. Then have students play notes on a keyboard to match a fast metronome speed and a slow speed.

To focus on BIM 6, ask: *Why does it make sense to call the fast beats small and the slow beats big?* [If the notes are faster, there is less time between them, so they are smaller.] *If you measured a song using fast beats and then measured it again using slow beats, what would happen?* [I would get different numbers because there would be more small fast beats than big slow beats.]

Appropriate Non-standard Units for Time

As with all other types of measurement, the key notion behind using non-standard units for measuring time is that the units must be uniform and appropriate. When students use tools for measuring in non-standard units, it is important for them to know that they must use the tools consistently to ensure uniformity. Very young children can use units such as claps or counting to measure time, but students will eventually realize that it is not possible to ensure uniformity with units that vary from person to person or even within the same person. Tools such as metronomes and sand timers provide these uniform units. We use shorter units to measure short durations or to increase precision, while longer units measure longer durations.

Estimating Time with Non-standard Units

It is valuable for students to learn not only to measure time, but also to estimate time measurements. You should give students many opportunities to estimate and then check how many time units are needed to accomplish a variety of tasks. As with other measurements, they can use activities with known durations as referents for estimating. For example, students might learn that a pendulum swings 7 times while they write their first name. They can use this information to estimate how many times the pendulum will swing as they write their full name or as they copy a sentence from the board.

Time: The Standard Units Stage

Standard Time Units

MINUTES (min)

Defining a Minute

The metric system defines a minute in terms of a second, but seconds are unfamiliar to students. It makes better sense to define a minute as either about the amount of time it takes to count to 60 slowly or as the time it takes the long hand on a clock to move one space over.

Minute Referents

To help students get a feel for one minute, you can use an estimating activity. First, have students watch as you time 1 minute. Then ask them to put their heads down, tell them to start timing, and have them raise their hands when they think 1 minute is over.

Other activities that help students establish personal referents include

- How many times can you write your name in 1 minute?
- How many times can you clap your hands in 1 minute?
- How high can you count in 1 minute?
- How many letters of the alphabet can you say in order in 1 minute?

HOURS (h)

Defining an Hour

An hour is 60 minutes.

Hour Referents

To help students establish personal referents for 1 hour, you can pose problems like these:

- What activity takes you about 1 hour to do?
- How far can you walk in 1 hour?

Familiar referents for 1 hour might include the amount of time students spend in math class, the length of a favourite TV show, or how long they have for lunch and recess.

DAYS, WEEKS, MONTHS, AND YEARS

- Students recognize that a day tells about the time from one sunrise to the next and that a week is 7 days.
- They think of months in terms of the calendar. It is about 30 days, a long length of time at the end of which you turn the page on the calendar and start recounting the days in the month.
- Students think of a year as the time it takes for all 12 months to pass. They may learn that it is about 365 days.

Telling Time to Measure Time

Students need to learn to tell time on both analog and digital clocks to be able to measure elapsed time. If they know the start time and the end time, they can figure out how much time has elapsed between those times.

Although the world is increasingly digital, students still need to be proficient at reading analog clocks because there are still many analog clocks in use. Also, understanding analog time helps students make sense of common time-related terms. The analog clock makes it apparent why you can read 9:15 as "a quarter past 9," as the minute hand has moved a quarter of the way around the clock. For students to relate to 9:15 on a digital clock as a quarter past the hour, they have to employ a considerable amount of number sense, realizing that 15 minutes is $\frac{1}{4}$ of 60 minutes.

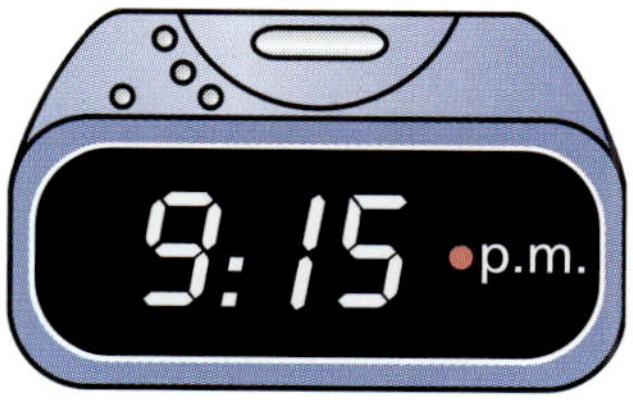

On an analog clock, a time such as "quarter past nine" makes sense visually because the minute hand is one quarter of the way around the clock face.

Most students can read the time on a digital clock on a superficial level, but unless they understand clock time, the times may not have much meaning for them. For example, a student might see 7:54 and read "seven fifty-four" without recognizing that this means that it is almost 8:00.

For students to make sense of digital times, they need to understand that

- there are 24 hours in a day that are numbered from 1 a.m. to 12 noon, and from 1 p.m. to 12 midnight (24-hour time is introduced later) (See A.M. and P.M. on **page 103**.)
- there are 60 minutes in 1 hour

Initially, students learn that there are two hands on an analog clock—the minute hand and the hour hand—and that the minute hand is longer. You may wish to have your students make their own paper-plate clocks with the hands labelled. (For teacher modelling, a geared clock is more suitable because it shows the movement of the hour hand as the minute hand moves.) Students usually have little difficulty telling time to the hour with an analog clock. Times that fall between the hours are more challenging until students have learned more about how these clocks work.

Teaching Idea 4.16

Suggest to students that they are playing a game with friends. They win if they are better at predicting how many times they can do a certain action in a given amount of time. Ask students to estimate how many times each can happen in the time it takes to empty a sand timer. Then check by doing the action:

- how many steps they can take
- how many jumping jacks they can do
- how many times a metronome beats at a specific setting

Action	My estimate	My result
Steps	50	90
Jumping jacks		

To focus on BIM 2, ask: *Does measuring in jumping jacks give you a better answer than measuring in steps? Explain your thinking.* [No; They are different answers but one isn't better than another.]

To focus on BIM 5, ask: *Why is it easier to compare the time for two students if we use the metronome beats instead of the steps?* [Metronome beats stay the same while we measure, but different students might take different size steps or one student might change the size of her steps while measuring the amount of time something takes.]

To focus on BIM 3, ask: *Why is it a good idea to estimate before you take a more exact measurement?* [If the measurement is very different from my estimate, then I know I should check my answer.]

Essential Notions about Analog Clocks

HOW THE HAND MOVEMENTS RELATE

As the minute hand travels around the clock, the hour hand moves much more slowly from one number to the next. So, except for time on the hour, the hour hand does not point directly at the number you need to say.

At 6:25, the hour hand points to a spot that is a bit less than halfway between 6 and 7.

CLOCK NUMBERS AND MINUTES

The numbers on a clock (1, 2, ..., 12) correspond to the hours, not to the minutes. For example, when the minute hand points to the 1, you say the time is "5 after," not "1 after."

To interpret the number of minutes after the hour, students must count by 5s according to the numbers (1, 2, ...) around the clock face.

DIFFERENT MEANINGS FOR DIFFERENT HANDS

We say different words when different hands point to the same number on the clock. Students need to be attentive to the hand lengths and the numbers.

6:00

12:30

When the hour hand points to the 6, we say "six," but when the minute hand points to the 6, we say "thirty."

Some teachers use clocks with an outer numbering system that shows the number of hours and minutes. This helps students associate the number of minutes after the hour with the position of the minute hand.

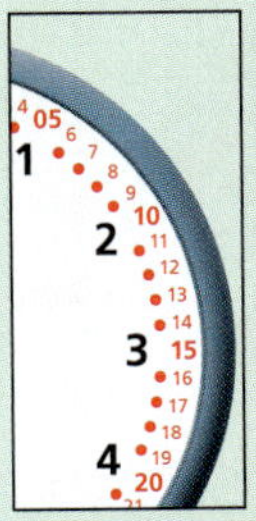

MULTIPLE NAMES FOR THE SAME TIME

If the time falls more than halfway through the hour, we can read it either as a number of minutes after one hour or as a number of minutes before the next hour.

You can report this time as "three forty" or as "twenty (minutes) to four."

Teaching Idea 4.17

Tell students that a movie starts at 2 p.m. and is 110 minutes long. Ask if they can be sure what time the movie ends. [Yes; It ends 110 minutes after 2 p.m.]

To focus on BIM 7, ask: *Can you be sure about the end time if the movie is advertised as being 20 songs long? Why?* [no; because songs are all different lengths]

Because of these complexities, teachers usually sequence work with analog time so students learn to tell time first to the hour, then to the half hour, to the quarter hour, to the nearest 5 minutes, and finally to the nearest minute. If a clock also has a second hand, telling time to the minute can be especially challenging because students need to observe the position of the second hand as well as the minute and hour hands.

Setting the time on a digital or analog clock is a different skill from simply reading the time. A sample setting time question might be, "On this clock face, draw what the hands of the clock look like at 3:15." Because most curriculum expectations/outcomes are about reading clocks, you should focus on this in your classroom activities. However, it is a useful life skill to be able to set a watch or clock. For some students, this will come later. Many interesting problems can be related to the position of hands on a clock without requiring students to actually set the time, for example, "At about what time are the hands of a clock opposite one another?"

A.M. and P.M.

Time is usually described using the 12-hour clock, in which the hours cycle from 1:00 in the morning until 12:00 noon and then repeat the cycle from 1:00 in the afternoon until midnight. Since having two times with the same name in a single day can be confusing, you can specify whether you mean morning or evening by using the acronyms *a.m.* and *p.m.* These stand for the Latin words *ante meridiem* and *post meridiem*, which mean "before noon" and "after noon," respectively. A good way to introduce the concept of a.m. and p.m. is to use a full-day timeline.

> **Teaching Idea 4.18**
>
> Tell students that you just finished reading the newspaper and the clock says 10:00.
>
> To focus on BILT 2, ask: *Why can you not be sure how long it took me to read the paper even though you know what the clock says?* [I need to know the start and the end time to measure time, and I only know the end time.]

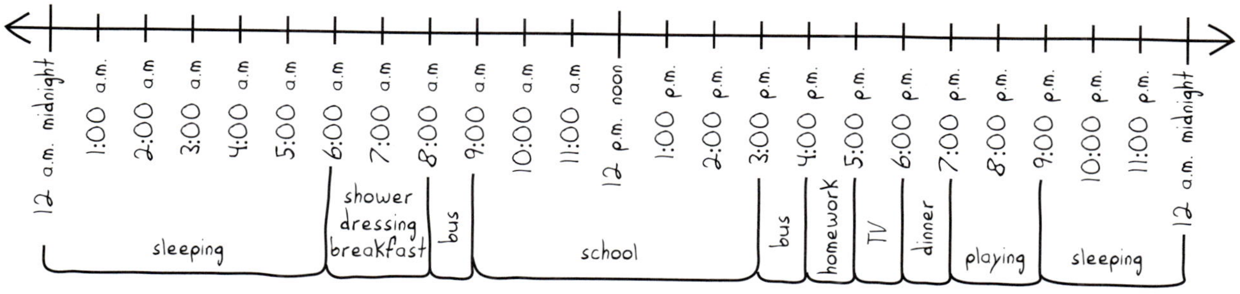

A full-day timeline can help students with a.m./p.m. notation. It clarifies the common confusion of 12 a.m. (midnight) and 12 p.m. (noon).

Measuring and Calculating Elapsed Time

As described above, once students can tell time, they can use a start time and an end time to calculate elapsed time. For young students, it makes sense if the times are simple, for example, both hours, both half hours, or both within the same hour: 6:30 to 6:45, but not 6:30 to 7:15.

Students might use a full-day timeline like the one above to help them visualize and figure out elapsed time longer than an hour. For times within an hour, they might use a minute timeline like the one below.

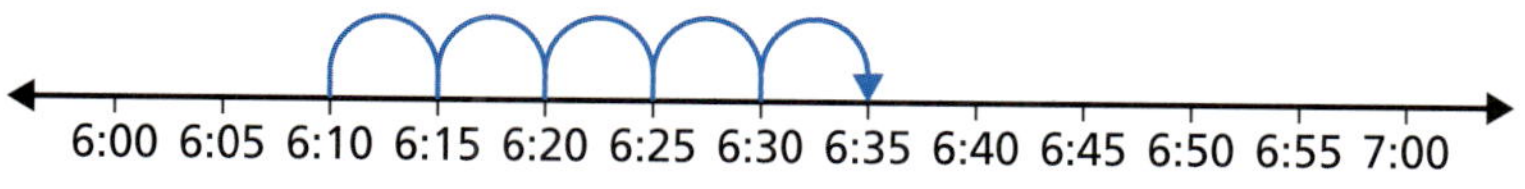

From 6:10 to 6:35 is 5, 10, 15, 20, 25, or 25 minutes.

Students can also use an analog clock with marks at each minute and count the number of minute hand moves.

> **Teaching Idea 4.19**
>
> Tell students that an event started at 1:00 and ended at 5:00. You want to know how long it took.
>
> To focus on BIM 6, ask: *Why would you measure the time in hours and not in minutes?* [The number is a lot easier to understand if I use hours. I can imagine 4 hours, but it's hard to understand how long 240 minutes is.]

Area

While length describes a one-dimensional attribute of a shape, area describes a two-dimensional attribute—the amount of two-dimensional (flat) space that a shape covers or that is required to cover the shape.

The teaching ideas in this part of the chapter will indicate which Big Idea(s) for Measurement (BIM) can be emphasized as well as whether the Big Idea for Area (BIA) can be emphasized.

BIG IDEA FOR AREA

1. Area tells how much material is required to cover a shape on the surface of an object. Shapes that look different can have the same area. (BIA 1)

Measuring Area

At the K to 3 level, students work mainly at the definition/comparison stage and sometimes at the stage of using non-standard units.

Introducing Concepts of Area in the First Two Stages

DEFINITION/COMPARISON STAGE	NON-STANDARD UNITS STAGE
	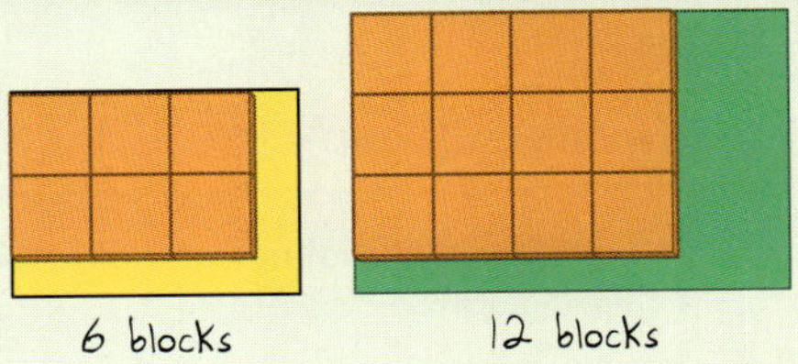
"The yellow rectangle is smaller because it doesn't cover all of the green rectangle."	"The yellow rectangle is smaller because I used 6 blocks to cover most of it and 12 blocks to cover most of the green rectangle."

Sometimes students confuse perimeter and area, so you need to emphasize that perimeter measures the line around the outside, while area measures the inside space. You might even put a visual like this up on display.

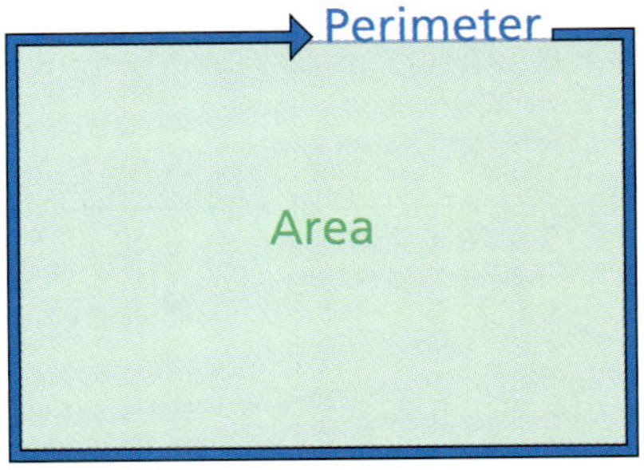

A visual reminder to differentiate area and perimeter

Area: The Definition/Comparison Stage

At this stage, students explore the notion that one item has more area than another if you can cover one shape with the other and there is still some part that is uncovered.

Some sample measuring tasks might be measuring

- two coins to see if a Canadian or an American quarter is bigger
- two pictures to see which will take up more space on the bulletin board
- a photograph to see how many will fit on a page in an album

Sometimes one shape does not fit completely on top of the other. When this happens, students may find they need to dissect one of the shapes in order to make a direct comparison. In the example shown here, a student tried unsuccessfully to fit the orange rectangle over the blue rectangle, and eventually discovered that cutting the orange rectangle in half made it easier to see that the blue rectangle was larger.

First try Second try Third try—dissecting and comparing

Sometimes students need to dissect one of the shapes to make a direct comparison.

In order for students to use this dissection method, they need to understand that the area of a shape does not change when it is dissected, as long as no parts are removed. This understanding is called conservation of area, and it develops at different times in different students.

Area: The Non-standard Units Stage

Because it is often difficult to compare the areas of two shapes unless one fits completely on top of the other, students move quickly to the notion of covering each shape with the same non-standard units to decide which of two shapes covers more area.

Appropriate Non-standard Units for Area

Non-standard area units are items of uniform size that you can use to tile a surface. To tile means to cover a surface with no gaps or overlapping. Suitable items include sheets of paper or newsprint (to cover big surfaces), pattern blocks, square tiles, attribute blocks, ink stamps, or even cookie cutters.

Students iterate a single area unit when they use a cookie cutter to measure the area of a piece of rolled out clay. They end up with a record of each iteration.

Teaching Idea 4.20

Explain to students that when companies place ads in a magazine, they pay according to how much space they take up on a page. Ask students to cut five ads out of a magazine or newspaper, including two ads that they think are about the same size. Ask them to order the ads from least area to greatest area, reminding them that area tells about how much surface something covers.

To focus on BIA 1 and BIM 1, ask: *Can pictures that are different shapes have the same area? How do you know?* [Yes; e.g., When I cut up the skinny picture, it covers the wide picture perfectly.]

Teaching Idea 4.21

Provide students with two sheets of paper that have the same area but that look different, such as a 10 × 12 blue rectangle and a 5 × 24 red rectangle. Tell them that each is the size of a rectangular pizza from a different company, but one company charges more than another. Ask them to decide which pizza should cost more. Then have them cover the papers with the same unit to test their prediction.

To focus on BIA 1 and BIM 1, ask: *How can you create another shape (pizza) with the same area as these two but that looks different?* [e.g., Cut the long thin shape in half and then put the halves together to make a longer, thinner shape.]

From time to time, you should give students a choice about which unit to use to measure an area. These situations help students build their understanding of area concepts and provide concrete experience that will support them when they move on to work with standard units. The examples that follow illustrate some of the area concepts students can develop as they work with various types of units.

Important Area Concepts

AREA UNITS THAT DO NOT TILE

Students should measure area with units that do not tile only during introductory work with area, when they are exploring why some units work well for measuring area and others do not. When students use units that do not tile, they learn that much of the area is not counted in the measurement, and that different students can get different results for the same area, depending on how they arrange the units.

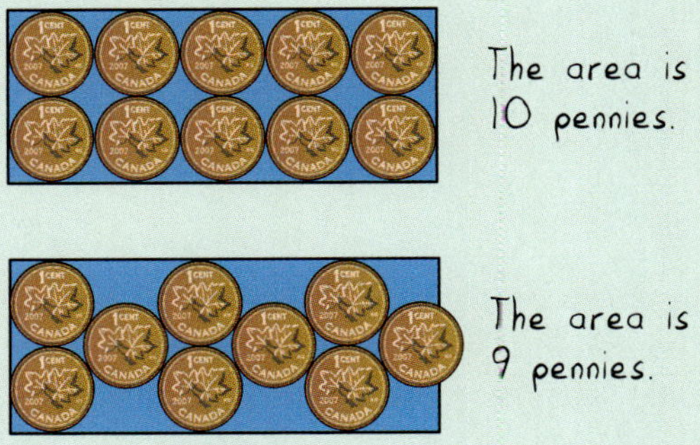

Circular items are not appropriate area units because you can get different results, depending on how you arrange the units.

AREA UNITS THAT DO NOT COMPLETELY COVER A SURFACE

Even when units do tile, the units may not cover some parts of a shape. Students must estimate the uncovered area. They can do this in different ways.

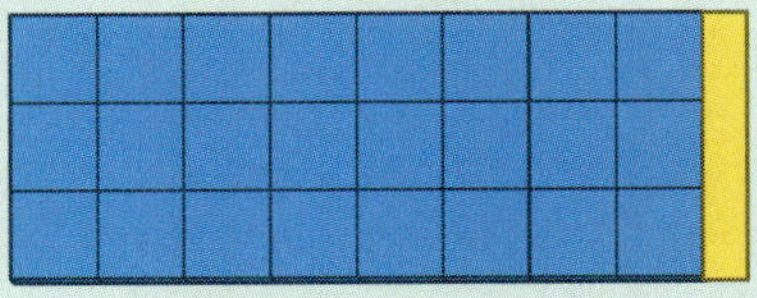

"That's 24 whole squares and about 1 more, so the yellow area underneath is about 25 squares."

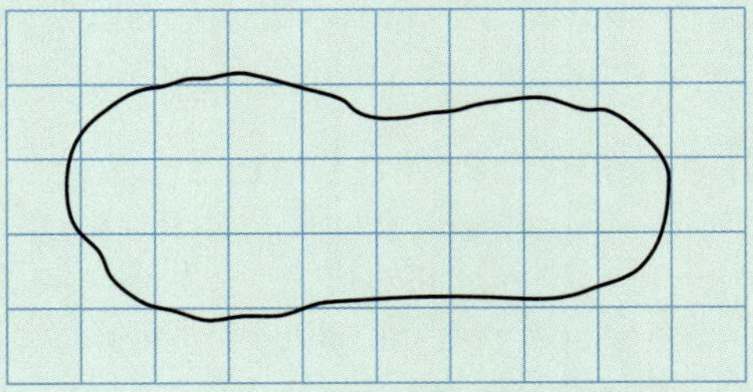

"There are 50 squares altogether and about 30 aren't covered, so the area of my shoe is about 50 − 30 = 20 squares."

USING NON-SQUARE AREA UNITS

Work with non-square units helps students learn that squares are not the only shapes you can use to measure area. When students use units that tile, they learn that it is important to fit the units together with no spaces between them. Otherwise, large amounts of area are not counted in the measurement.

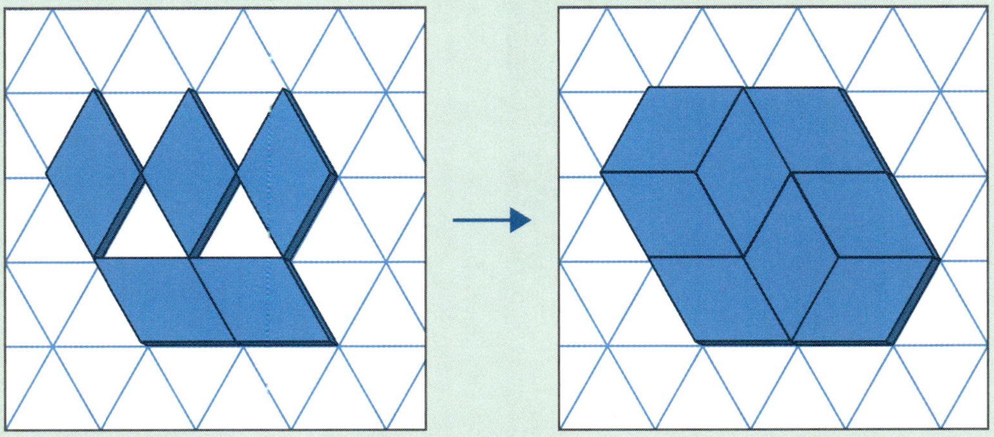

It helps if the surface to be covered is a different colour from the units being used so that it is easy to see any uncovered surface.

Strategies for Measuring and Estimating with Non-standard Units

Students need to recognize that there is never only one way to measure an area. For example, to determine or estimate the area of the red triangle below, students at the non-standard unit stage could

- cover the triangle with concrete units, such as triangle pattern blocks, and count how many units it takes to cover the triangle, estimating as necessary to account for area not covered by the blocks
- place a transparent square grid over the triangle and estimate how many whole grid squares cover the shape; they might estimate with partial squares by counting only the squares where half or more of the square is within the shape

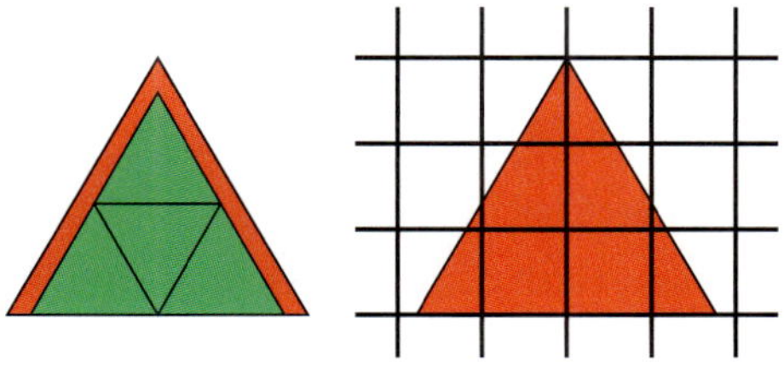

The red triangle's area is a bit bigger than 4 green block triangles. Or, it's about 6 squares.

Measuring the Same Area with Different Non-standard Units

Students need opportunities to measure the same area using different units to help them discover that they can describe the area in different ways. They will discover that as the size of the unit increases, the number of units decreases. For example, an area that covers 49 squares on a centimetre grid can also be covered with only about 8 square pattern blocks. Experiences like this also help students discover that it is more efficient to use larger units to measure larger areas and to use smaller units to measure smaller areas.

Measuring Different Areas with the Same Non-standard Unit

Not only do students benefit from opportunities to explore area by covering the same shape with different units, they also benefit from using the same unit to measure different areas. A common unit is especially useful when students are comparing the areas.

Teaching Idea 4.22

Give students a shape that was made by outlining four yellow pattern blocks.

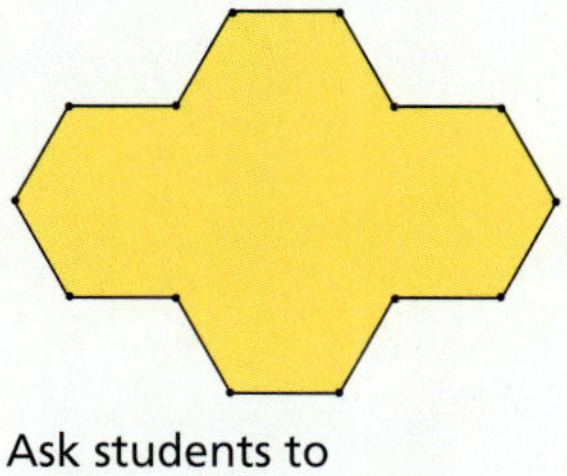

Ask students to

- estimate the area of this shape using yellow pattern blocks and then measure to check their estimate
- repeat with the red, blue, and green pattern blocks

To focus on BIM 6, ask: *Why were the numbers of area units different for the red blocks and the blue blocks?* [The red blocks are bigger, so I don't need as many of them.]

To focus on BIM 4, ask: *How can knowing the area in yellow blocks help you predict the area in blue blocks?* [I know 3 blue blocks cover 1 yellow block. I used 4 yellow blocks, so I added 3 four times: 3 + 3 + 3 + 3 = 12.]

To focus on BIM 5, ask: *If you want to compare the area of two shapes using blocks, why does it make sense to use the same blocks for both?* [It is easier to compare if I use the same units.]

Capacity and Mass

We use capacity to describe the amount of space inside an object that is a container, such as a cup, a pail, a swimming pool, or even a room. Mass measures the amount of matter contained in an object. Capacity and mass are both measures of 3-D objects.

BIG IDEAS FOR CAPACITY AND MASS

The teaching ideas in this part of the chapter will indicate which Big Idea(s) for Measurement (BIM) and which Big Idea(s) for Capacity and Mass (BICM) can be emphasized.

1. The capacity of an object tells how much it will hold as a container. A big object can have a smaller capacity than a smaller object, depending on thickness of the walls or the dimensions of the containers. (BICM 1)
2. The mass of an object tells how heavy it is. A big object can have a smaller mass than a smaller object. (BICM 2)

Capacity

You can measure capacity indirectly using a liquid or a pourable substance like sand. The amount of liquid or pourable substance that fits inside the container is a measure of the container's capacity.

Although students at the K to 3 level do not study volume formally, you need to be aware of the difference between volume and capacity so that you do not inadvertently contribute to the common confusion between the two types of measurement. Volume is a measure of the size of an object—how much three-dimensional space it takes up—and is measured using non-standard volume units, such as cubes, and standard units, such as cm^3 and m^3. Capacity is the amount of three-dimensional space inside an object, assuming it is a container, and is measured using non-standard units, such as cupfuls, or standard units, such as L and mL. When we use a capacity measure to describe a liquid or pourable substance, for example, 1 L of milk, we mean enough milk to fill a 1 L container.

You can introduce ideas about capacity using the same three stages as for other measurement concepts—definition/comparison, non-standard units, and standard units. As with length, the second stage is usually introduced fairly soon after the first stage.

Teaching Idea 4.23

Ask students to think of situations where they might want to know if a pail or mixing bowl holds more than another one. Then provide a container marked with an X. Give students several other containers of varying capacities and shapes, including some that hold about the same amount as Container X. Ask students to estimate how each container might compare with Container X (they could line them up in order), and then measure to check, recording the results.

To focus on BICM 1 ask: *What did you look at to help you decide how the capacities compared?* [I looked at how deep and wide the inside of each container was.]

Introducing Concepts of Capacity in Three Stages

DEFINITION/COMPARISON

"The blue container holds more because there's still room in it after I pour in all the sand from the red container."

NON-STANDARD UNITS	STANDARD UNITS

"The blue container holds more because it holds 5 scoops of sand and the red one only holds 3 scoops."	"The blue container has a greater capacity because it holds 500 mL and the red one only holds 300 mL."

Capacity: The Definition/Comparison Stage

At this stage, students explore the idea that you can compare the capacity of two containers by filling one of the containers with a pourable material, such as water, sand, rice, or small peas, and then pouring the contents into the other container. As they compare, students discover that the capacity of a container is determined not only by its height, but also by the area of its base and the configuration of its sides. You should give students opportunities to compare containers that differ in only one dimension as well as containers that differ in more than one way.

Comparing Containers

When only one dimension differs ...	When more than one dimension differs ...
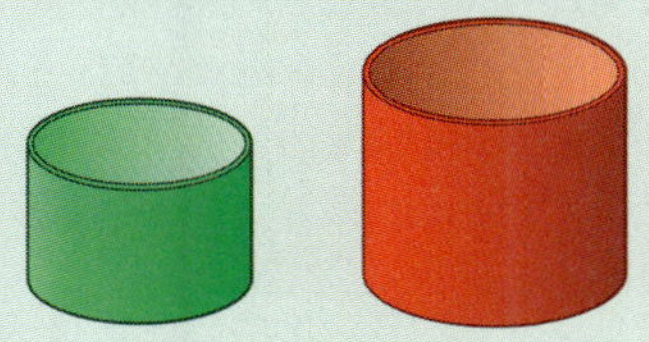	
... it is easy to see at a glance which holds more (as long as the container walls are of similar thickness).	... it can be difficult to compare visually. In these situations, students can pour the contents of one into the other to compare.

In the course of measuring capacity, students will use a variety of measurement terms, such as *holds more*, *holds less*, *full*, and *empty*.

Appropriate activities for students at the definition/comparison stage include comparing the capacities of empty drink containers (perhaps collected from students' lunches), comparing the capacities of cooking pots, or deciding what size bags to use for a class popcorn sale. Students should begin by comparing only two containers, but they will soon be ready for the greater challenge of putting three or more containers in order. A sand or water table is a good place for younger students to compare capacities.

Capacity: The Non-standard Units Stage

At this stage, students learn to measure capacity with non-standard units such as tea cups and soup spoons. Non-standard units make it possible to assign a number to a container's capacity, so students can record the capacities of containers to compare them without having to refill them repeatedly.

Non-standard units also help students develop referents they can use for estimating. For this reason, students need many opportunities to measure using familiar units, such as water bottles, yogurt cups, cereal bowls, and spoons. A student who knows, for example, that a container holds 4 soup spoons of sand has a good feel for the capacity of the container.

Teaching Idea 4.24

Many students assume that a taller object automatically holds more because it is "bigger." Show students two glasses, one a tall glass full of water and the other an empty shorter, wider glass with a greater capacity.

To focus on BIM 1 ask: *Do you think it is possible for the shorter glass to hold more water? How could that be?* [Yes; It's wider.] Then empty the water from the tall glass into the short glass to check.

Teaching Idea 4.25

Use linking cubes to create two structures like these. Notice that the outside dimensions are the same, but the inside dimensions differ.

To focus on BICM 1 and BIM 1, ask: *Both structures are the same size. Does that mean they hold the same amount? Explain your thinking.* [No; They are the same size on the outside but not on the inside.] Then ask students to build two cube structures that are the same size but with different capacities.

At this stage, students can use multiple units to measure capacity or they can use the same unit over and over. It may be more practical to use one unit repeatedly. You should emphasize the need to fill a unit uniformly level to the brim, but not farther.

Teaching Idea 4.26

Show students a pail that is about $\frac{1}{3}$ full of apples. Ask them to estimate how many apples it would take to fill the pail.

To focus on BIM 4, ask: *How did you use what you already knew to help you estimate?* [There were about 10 apples already, so I estimated there was room for three times that many, or 30.]

To focus on BIM 3, ask: *Why might it be useful to estimate how many apples a pail could hold?* [e.g., If I went to pick apples and I knew how many apples I wanted, I'd know how many pails to take.]

To focus on BIM 6, ask: *What if the apples were smaller? How would that change the number the pail could hold?* [The pail would hold more apples.]

To focus on BIM 5, ask: *Suppose you know that another pail holds 35 apples. Can you be sure whether it holds more than the first pail? Why?* [No; The apples may not be the same size.]

Teaching Idea 4.27

Provide a small scoop, a graduated measuring cup like the one shown at the left, and a small bowl full of sand.

To focus on BIM 2, ask: *What are some different ways you can measure how much rice my bowl holds?* [e.g., I can pour the sand into the measuring cup and see how high it goes or I can see how many scoops of sand there are.]

Appropriate Non-standard Units for Capacity

Non-standard units for measuring capacity are handy everyday items of uniform size, such as spoons, cups, and small containers. You can fill them with water or with a pourable solid such as sand or rice.

If students are measuring boxes, cubes make good non-standard units because they pack (i.e., fill without gaps or overlap). In the early stages of work with definition/comparison, you may wish to let students experiment with units that suit a particular context but do not pack. For example, students could figure out how many apples will fit in a basket or how many pennies or jelly beans will fit in a jar. However, whenever you use these materials, you should discuss with students afterwards that they did not measure some of the space in the container because the units did not fit tightly together. As a result, it might not be good to use these units if you want to compare the capacities of two containers that are similar in size.

It is important to note that, although units such as measuring cups, teaspoons, and tablespoons are in fact standard imperial units, they may be used as non-standard units in this context.

Estimating Capacity with Non-standard Units

In the definition/comparison stage, students estimate to determine which container holds more. Non-standard units make it possible to assign a number to an object's capacity, allowing students to record the capacities of objects so they can compare them without having to measure over and over.

As with other types of measurement, it helps students to compare an unknown amount with an amount they know. For example, a student at this stage might determine that it takes 5 full tea cups to fill a container, and then estimate how many larger cups or mugs it will take to fill the same container. At this stage, you can fill a container partially, for example, to approximately $\frac{1}{2}$ or $\frac{3}{4}$ of its capacity, and then ask students, "About how many units altogether will it take to fill the container?"

Making Non-standard Measuring Tools

A good way to help students understand the importance of uniformity in non-standard units is to let them measure out equal amounts in order to make their own measuring tools. For example, students can make a measuring cup from a large transparent cup or glass (preferably cylindrical rather than tapered) by filling it repeatedly from a smaller container and marking the side of the cup to show the level after each new measure is added. Students can then use the cup to measure the capacity of various bowls.

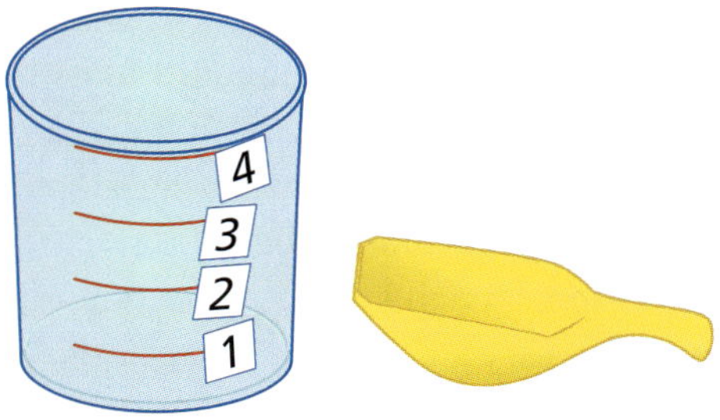

A student-made non-standard unit measuring cup

Capacity: The Standard Units Stage

Students can start using standard units when they realize that, rather than measuring capacity in non-standard units, which can mean different things to different people, they can use units that everyone will understand.

Introducing Standard Units of Capacity

We usually introduce the litre early in work with capacity because litre containers are a familiar part of everyday life; products such as milk and ice cream often come in litre containers. This familiarity also helps students understand that litre containers can look different; a milk carton can look quite different from an ice cream tub.

THE LITRE (L)

Defining a Litre
A litre is the amount of liquid it takes to fill a cubic container that is 10 cm by 10 cm by 10 cm.

Litre Referents
Aside from familiar referents such as 1 L milk cartons and water bottles, students can also refer to a thousand cube from a base ten block set.

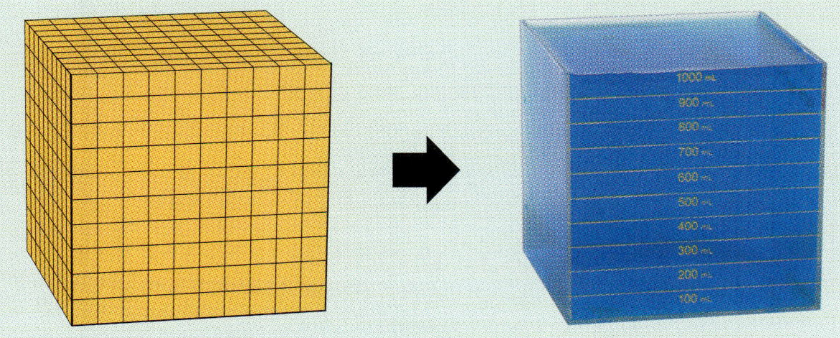

If a thousand base ten cube were a container, its capacity would be 1 L.

Teaching Idea 4.28

Tell students that you asked your mother to prepare soup for a party you were having. You told her you needed enough for 6 bowls. When you started to serve the soup, you discovered there was not enough soup.

To focus on BIM 7, ask: *How could that have happened?* [Your mother used a smaller bowl.] *What could I have told my mother so there would be no mistakes?* [exactly what size bowl you were using or how many litres of soup to make]

Mass

In Canada, we generally teach children to use the term mass, rather than weight, to describe the heaviness or lightness of an object. These are the differences between mass and weight:

- Mass measures the amount of matter contained in an object, while weight is a measure of force—the combined effect of mass and gravity.
- The mass of an object is measured by using a balance to compare it to a known amount of matter, but weight is measured on a scale. (Scales are affected by the pull of gravity and, as such, are used to measure the weight of an item; whereas the effect of gravity is negated when using a pan balance since the items on both sides of the balance are affected in the same way.)
- The mass of an object does not change when location changes, but weight can change with a change in location.

In everyday informal use, the terms mass and weight are interchangeable. In fact, we often tell a student to weigh an object to determine its mass. However, some more advanced scientific and technological applications require a careful distinction between the two terms.

You can introduce ideas about mass using the same three stages as for other measurement concepts—definition/comparison, non-standard units, and standard units. Unlike the teaching about length, here it is critical to use tools such as pan balances even at the definition/comparison stage.

Introducing Concepts of Mass in Three Stages

DEFINITION/COMPARISON

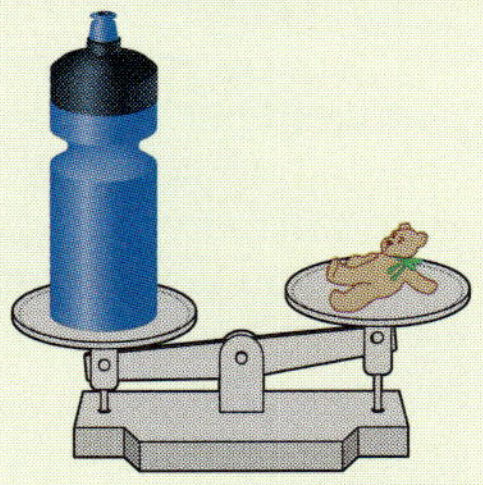

"The water bottle is heavier because the pan went down on that side."

NON-STANDARD UNITS

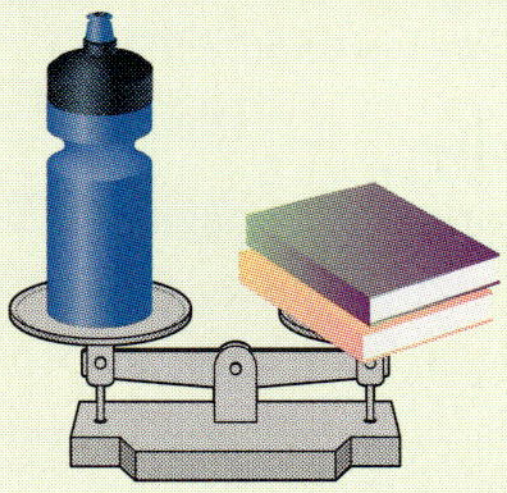

"The water bottle has a mass of 2 books."

STANDARD UNITS

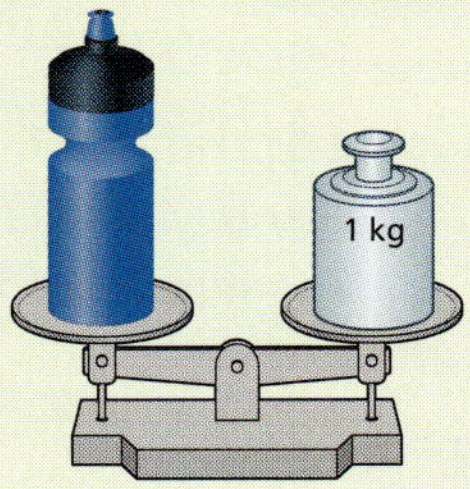

"The water bottle has a mass of 1 kilogram."

Teaching Idea 4.29

Ask students to find a small but heavy item. Then ask them to find a larger but lighter item. Have them test their choices using a pan balance.

To focus on BICM 2 and BIM 1, ask: *Why can you not just look at the items to tell which is heavier?* [If something is made of heavy stuff, it can be smaller but heavier than something big made out of light stuff.]

Mass: The Definition/Comparison Stage

At this stage, have students use a pan balance to compare the masses of two objects. The balance acts like a teeter-totter; the object whose mass is greater pulls its side of the balance down (as shown above for definition/comparison).

Activities for students in the definition/comparison stage might include comparing the masses of different types of fruit or hunting for objects that have less mass (or more mass) than a given object.

Students at this stage may find it difficult to estimate which of two objects has more mass, since they cannot rely on size. Sometimes a small object has a greater mass than a large object. Therefore, students need to be able to handle the materials they are comparing. To develop their ability to estimate in order to make mass comparisons, give students many opportunities to handle and compare a wide range of materials.

Another way to help students learn to estimate mass is to encourage them to use their own bodies as pan balances. Have students take an item in each hand, stretch out their arms, and try to feel which item is heavier. Ask them to move their arms to show how a pan balance would look if the items in their hands were on the balance.

Using a Pan Balance to Compare Mass Directly

Students can make their own pan balances, for example, by using string and a paper clip to connect a paper cup to each end of a coat hanger. It is also practical to have some commercial models on hand.

Before using a pan balance, level it by making sure the empty pans balance exactly. Pan balances tend to slip out of balance when not levelled regularly. An unlevelled balance will result in inaccurate measurements and may foster misconceptions about mass. Most classroom pan balances are suited for a limited range of masses, usually up to 1 kg or 2 kg.

A STUDENT-MADE PAN BALANCE

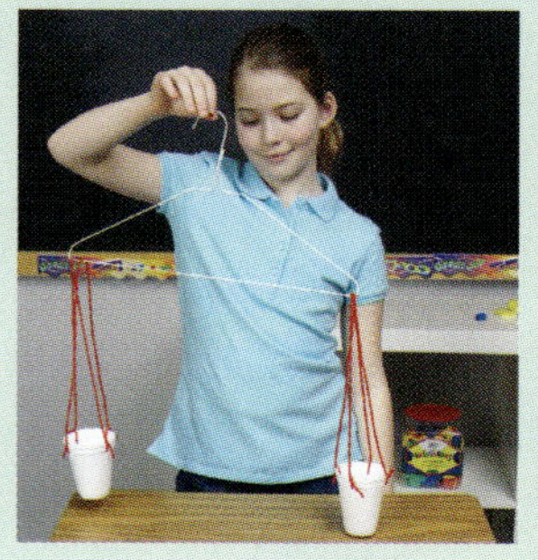

A COMMERCIAL PAN BALANCE

Some commercial balances have a flat plate on each side, making it easy to see the objects being measured. Others have removable containers, often clear, which allow you to balance liquids or small items that might otherwise roll away.

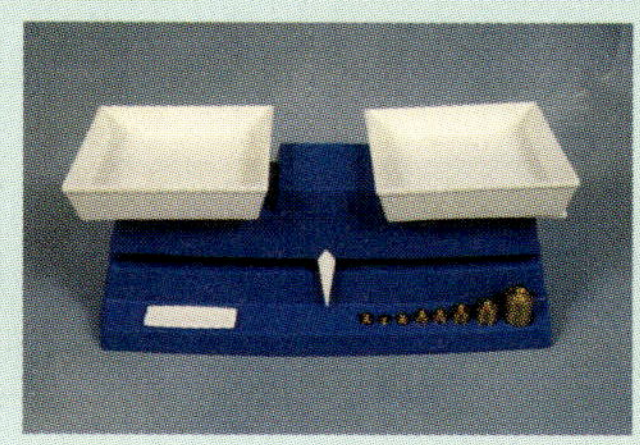

Many students benefit from activities that help them verify that breaking up a mass does not change its total "heaviness." You can ask students to create two balls of modelling clay with the same mass, verified using a pan balance. Then ask them to break up one ball into several small balls and predict whether all of the small balls together will balance the larger ball. They can use the balance to verify that the mass did not change.

Mass: The Non-standard Units Stage

Non-standard units make it possible to assign a number to an object's mass, allowing students to record and compare the masses of many objects without having to measure over and over.

To measure an object's mass with non-standard units, place the object on one side of a pan balance and then count how many uniform units you must place on the other side to make the pans balance. For example, it might take 8 blocks to balance the mass of a pet rock.

Appropriate Non-standard Units for Mass

Students can measure mass with non-standard units such as marbles, coins, linking cubes, base ten blocks, ball bearings, unused erasers, or other small counters. For practical reasons, make sure the unit has a reasonable amount of mass—not too much or too little. For example, if you use table tennis balls, you will need a lot of them to measure the mass of even a light object like an apple, making them an impractical unit. A unit with too much mass, such as a book, does not allow for much precision in measuring mass.

Measuring Mass with Different Non-standard Units

Work with different non-standard units allows students to explore the big idea that using a bigger unit to measure mass results in a smaller number of units. You should give students opportunities to balance a particular object with several different types of units, recording the results each time. Work with different units also helps students develop personal referents that they can use to estimate masses. For example, you can ask students to measure the mass of a tennis ball using non-standard mass units like pennies, cubes, and pattern blocks.

I measured the tennis ball

What I Used	My Guess	How Many?
pennies	15	21

Students can base each estimate both on the previous measure and on how the next unit compares to the previous unit.

Teaching Idea 4.30

Ask students to measure the mass of a tennis ball first using pennies and then using small plastic counters.

To focus on BIM 6, ask: *Did the mass of the tennis ball change? How do you know?* [No; It's the same ball each time.] *Why was the number of pennies needed different than the number of counters?* [Counters are lighter than pennies so I need more counters to measure the same mass.]

To focus on BIM 5, ask: *Suppose another ball had a mass of 28 pennies instead of 21. Can you be sure it is heavier? Why?* [Yes; A penny always has the same mass, so 28 pennies is heavier than 21 pennies.]

Measuring Different Items with the Same Non-standard Unit

Students need opportunities to balance different items against a common unit. For example, students could use linking cubes to balance a series of small toys. Using the same non-standard unit for each item allows students to compare masses to see which item has a greater mass, and also to determine the amount of difference.

Activities that provide a context for measuring different items with the same unit include hunting for items with a particular mass, such as "10 marbles," or putting a series of objects in order according to mass.

Teaching Idea 4.31

Provide various small units for measuring mass, such as square pattern blocks, pennies, and linking cubes. Have students use different units to estimate and then measure the mass of a small ball.

To focus on BIM 3, ask: *Why might it be useful to estimate the mass of a ball?* [e.g., to find out how many balls I could carry]

To focus on BIM 4, ask: *Which unit made it easiest for you to estimate? Why?* [e.g., Linking cubes; I already had a pretty good idea of how heavy 10 linking cubes was, so I compared that in my mind to how heavy the ball was.]

Estimating Mass with Non-standard Units

In the definition/comparison stage, students estimated to determine which of two objects had more mass. Now students are able to estimate how many units it might take to balance a particular object. As with other types of measurement, it helps to compare unknown amounts with amounts they know. For example, a student at this stage might determine that it takes 10 linking cubes to balance a small object, and then estimate how many linking cubes it would take to balance a larger object.

It is more difficult to estimate the mass of an object than to estimate other measures because there is no direct relationship between the object's size and its mass. To estimate size (length, area, capacity, or volume), students can visualize how many units will fit along or inside the item being measured. With mass, students need to think about how the unit feels. It is not easy to feel one unit and then imagine what two units, or ten units, might feel like.

Mass: The Standard Units Stage

Students can start using standard units when they realize that, rather than measuring mass in non-standard units, which can mean different things to different people, they can use units that everyone will understand.

Introducing Standard Units of Mass

The kilogram is the first standard unit of mass that students usually encounter. Give students the opportunity to make a kilogram mass of their own. For example, they might fill a plastic container with sand until it exactly balances a 1 kg mass on a pan balance, or they can fill a 1 L milk carton with water.

The first standard metric unit students meet is the kilogram rather than the gram because a gram is simply too light to mean much to a young student. For example, it is meaningful to distinguish between items with a mass of 2 kg or 3 kg, but it is not really meaningful to distinguish between an item with a mass of 2 g and an item with a mass of 3 g.

The Kilogram—A Standard Metric Unit for Mass

DEFINING A KILOGRAM	KILOGRAM REFERENTS	MEASURING TOOLS FOR KILOGRAMS
One kilogram is a mass equal to the mass of 1 dm^3 of water at 4°C. 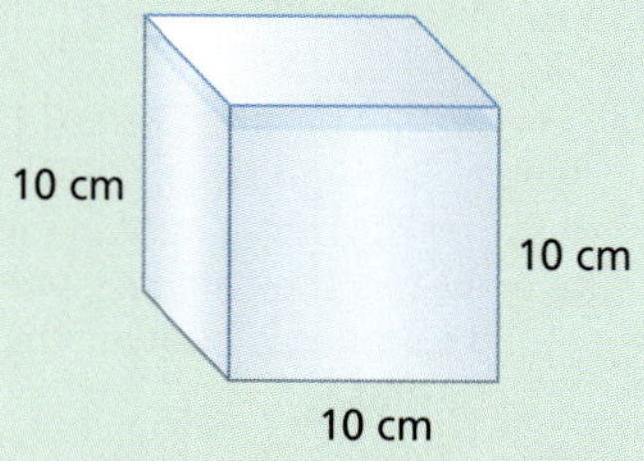 A 1 dm^3 cube of water	A kilogram is the mass of 1 L of water or milk. 	A pan balance is the preferred tool for measuring mass (see **p. 113**). However, since classroom pan balances seldom have the capacity to measure beyond 1 kg or 2 kg, you may need to use a scale to measure larger masses in kilograms. Because scales measure weight, not mass, scale measures may differ slightly from mass measurements depending on where, in relation to sea level, the scale is being used. At sea level, weight and mass are about the same, so weight is a good indicator of mass. Many scales have a feature that allows you to adjust for differences in altitude.

Chapter 5

Data and Probability

Organizing and Collecting Data

Collecting data satisfies human curiosity, whether it is to find out how many pets classmates have or to learn what TV shows your friends prefer. Organization is often the key to collecting data efficiently. Students in K to 3 can learn about themselves and their world and solve real problems by organizing and collecting data.

A knowledge of big ideas can help teachers choose, shape, and create tasks, and use questioning to help students make powerful connections.

BIG IDEAS FOR ORGANIZING AND COLLECTING DATA

1. When working with data, you usually organize or classify the data into meaningful categories. (BIOCD 1)
2. There are always many ways to sort or organize data. (BIOCD 2)
3. To collect good first-hand data, you must decide what collection method is most suitable and how best to pose any questions required to collect the data. (BIOCD 3)

The teaching ideas in this part of the chapter will indicate which Big Idea(s) for Organizing and Collecting Data (BIOCD) can be emphasized.

Organizing Data

Sorting and Classifying

Sorting is the physical act of grouping objects according to shared characteristics. *Classifying* is the process of differentiating among the groups by giving each group or category a name. Students' early work in data management includes sorting and classifying because many concepts and skills involved in sorting and classifying are fundamental to data management.

Many of the concepts and skills involved in sorting and classifying are fundamental to data management.

Students must be able to sort and classify, or organize and categorize data in order to record the results of data collection in an effective way. For example, if students collect data about their siblings, they can use different classifications or categories to sort or organize the data. They might choose categories like number of siblings, gender of siblings, or age of siblings, depending on the problem they need to solve.

Students need frequent practice with sorting and classifying objects. You can have them do sorting and classifying activities using everyday materials, such as cups, or with materials created specifically for sorting, such as the attribute materials and animal cards shown on the next page.

Teaching Idea 5.1

Ask students to sort a set of attribute blocks in as many ways as they can. Talk about the different possible categories (size, colour, shape, and thickness). Then ask students to create a one-difference train with the attribute blocks. Each block in the train must differ from the block before it and the block after it with respect to one attribute (e.g., a different colour or a different shape). For example, the block train below changes from a circle to a triangle (shape), then from blue to yellow (colour), and then from large to small (size).

To focus on BIOCD 2, ask: *Why was it useful to sort the blocks in different ways before you created your train?* [It showed me all the ways the blocks could be different so I was able to create a train that changed in a different way each time.]

Students should normally sort or classify for a purpose. For example, they might want to know what types of snacks to suggest for the school canteen, to organize materials for a class cupboard, or they might want to sort to show that different, but equivalent, patterns can be created with the same materials, for example, an ABB pattern based on shape and another based on colour.

Sample Sorting Materials

EVERYDAY OBJECTS

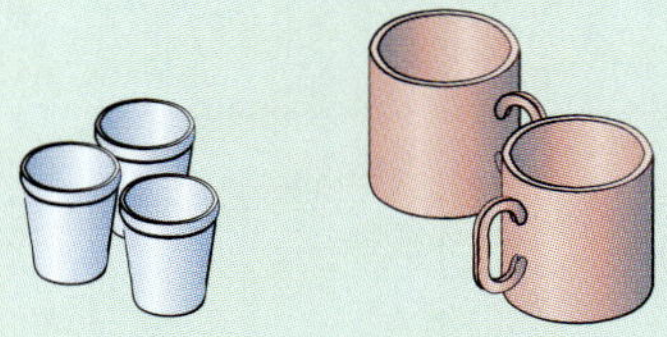

You can sort and classify using everyday materials. For example, the cups are sorted into two groups by size.

ATTRIBUTE BLOCKS

Attribute blocks are designed for sorting and classifying—there are five shapes in two sizes, two thicknesses, and three colours.

ANIMAL SORTING CARDS

The animal sorting cards are designed to be sorted by colour, type of animal, size, and shirt pattern. Each next card in this one-difference train differs by one attribute.

Teaching Idea 5.2

Students sometimes find it easier to recognize attributes if they are comparing objects. For example, students might compare a pencil and a ruler, listing everything they observe or notice about each object.

To focus on BIOCD 2, ask: *How can you use your descriptions to show that the objects are alike?* [e.g., Both are straight and both are used in school.] *How can you use your descriptions to show how the objects are different?* [e.g., Rulers have numbers all along and pencils don't; pencils are sort of round and rulers are flat.]

Recognizing Attributes

Before they sort and classify objects, students need to understand that every object has many attributes. Usually, activities that build this understanding are exploratory in nature—you display an object and ask students to describe it. For example, you can show students a T-shirt and ask them to describe it.

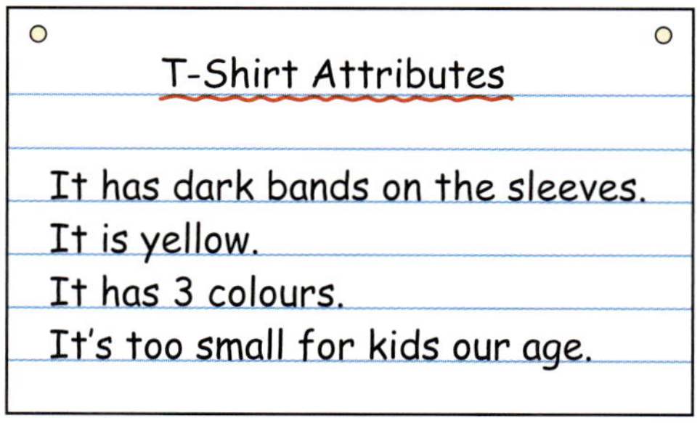

Brainstorming attributes of a T-shirt

Attributes and Characteristics

Some of the terminology associated with sorting can be confusing. People sometimes use the terms *attribute* and *characteristic* interchangeably. An attribute is a way to compare objects (e.g., by colour), while a characteristic describes how the attribute is reflected in a particular object (e.g., red, blue, green). You need to understand what these words mean so that you can model appropriate language, but you should not assess students on their ability to make the distinction. The chart below provides some more examples of attributes and possible characteristics for each.

ATTRIBUTE	POSSIBLE CHARACTERISTICS
General Attributes	
colour	red, yellow, blue
texture	rough, smooth, bumpy
Measurement Attributes	
size—length	long, short
size—volume	big, small
mass	heavy, light
Geometric Attributes	
how it moves	rolls, stacks
number of edges	5, 6, 7

Describing a Sorting Rule

Students need to practise recognizing and verbalizing a sorting rule for a pre-sorted set of objects. For example, students might try to figure out how the three children on the left below are different from the two on the right. They might describe the sorting rule as "students with orange shirts (or shoes)" (as opposed to those who do not have an orange shirt on). They could then re-sort using a different sorting rule, such as "students wearing shorts" (as opposed to those not wearing shorts, or those wearing pants).

Students with orange shirts

Sorting by shirt colour

Not only should students have opportunities to identify sorting rules, they should also create their own sorting rules and then sort according to those sorting rules or other students' rules.

It would be informative for students to see that sometimes two different sorting rules result in the same sort. For example,

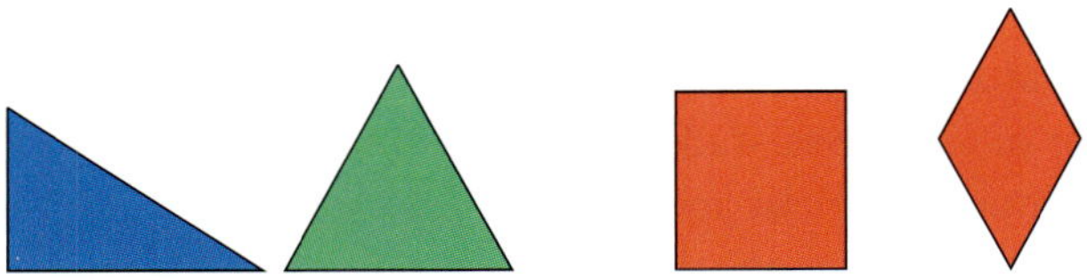

Rule 1: Shapes that are red
Rule 2: Shapes that are triangles

Teaching Idea 5.3

Suggest to students that they've decided to sort a set of toys into different bins to make it easier to find the ones they want. Provide a diverse set of small toy cars for students to sort by the attribute colour.

To focus on BIOCD 2, ask: *How else could you sort the cars?* [e.g., by size] *Suppose these were dolls instead of cars. Do you think there would be more than one way to sort them? Explain your thinking.* [Yes; e.g., I could sort by size: small and big dolls, or I could sort by the colour of their clothes.]

Other times, the same sorting rule might seem different, but leads to the same result. For example, asking students to sort shapes using the rule "shapes that are triangles" is no different from using the rule "shapes that have three sides."

Sorting by One Attribute

When sorting by one attribute, students sometimes apply a sorting rule by moving items together or by grouping them if they are alike. For example,

Sorting by one attribute—colour

If you ask students to sort in order to find all the red shapes, they must place the objects that are not red outside of the sorting circle marked red.

Shapes that are red

Sorting by one attribute—colour: red and not red

It is important to use concrete organizational tools to help younger students sort. For example, you might provide two (preferably transparent) containers labelled "red" and "not red" into which students can place the shapes they are sorting. Students can also use sorting mats or plates, or they can enclose items using yarn or Hula Hoops.

Some students have difficulty dealing with objects that do not possess the characteristic that is the focus of the sort. In the example above, they may find it easy to put the red objects together, but may be uncomfortable not having a name for the other objects. Rather than using the phrase "not red," they may come up with another name, such as "blue or yellow." Either name is acceptable.

At the early stages of development, especially preschool to Grade 1, students have an easier time sorting by the attributes of colour and size. Colour tends to be the easiest attribute for children to use in sorting. For students who continue to struggle in Grades 1 to 3, intervention strategies could use these "easier" attributes, before applying the same concepts to more complex attributes.

Sorting by Multiple Attributes

In a number of situations, students may need to sort in terms of two attributes, for example, by colour and by size. They might use a concrete or pictorial sorting chart like the one shown at the top of the next page.

Teaching Idea | 5.4

Propose a situation where students in a class are wondering whether short hair or long hair is more popular among 7-year-olds and suggest that they begin by finding out about their own class first. Ask all the students with short hair to stand in one part of the room. Then ask the class how they might describe the group of students that are still seated. [e.g., students with medium-length hair and long hair]

To focus on BIOCD 2, ask: *If I had asked all the students with medium-length and long hair to stand in one part of the room, would I have ended up with the same groups? How do you know?* [Yes; There would still be one group with short hair and one group with everyone else.] *What if I had asked all the students with long hair to stand in one part of the room? Explain.* [No; The students with short hair would be in a group with students who have medium-length hair.]

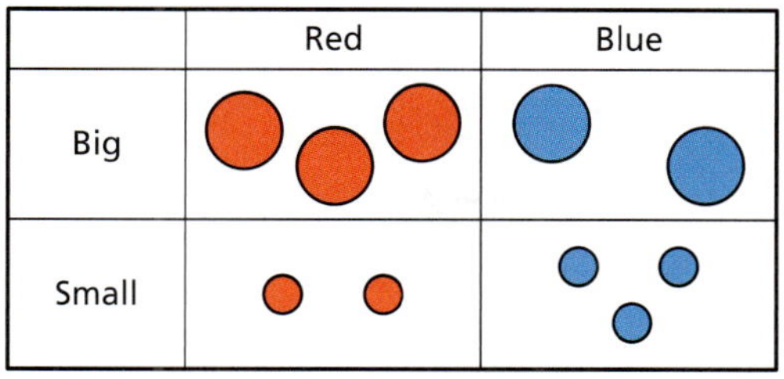

Sorting by two attributes—colour and size

Venn Diagrams

Venn diagrams are a commonly used sorting tool. Each circle contains all the objects with a particular characteristic. The name of the sorting rule is shown as a label for each circle. The number of circles depends on the number of characteristics or attributes you use to sort the items. There could be one circle (i.e., one sorting rule) or several circles (i.e., several sorting rules). Whether the circles overlap depends on whether one or more items in the set have multiple characteristics. The large rectangle that encloses the circle or circles represents the *universe*, or the whole set of objects you are sorting. Every object you sort has a place in the diagram: inside a circle, in an overlapping area, or outside the circles but inside the rectangle.

Every object you sort has a place in a Venn diagram: inside a circle, in an overlapping area, or outside the circles but inside the rectangle.

When you introduce students to Venn diagrams, you can use Hula Hoops or large rings of string or yarn as concrete Venn diagram circles. Initially, students can sort actual objects or pictures of the objects into these hoops. Later, they can draw the circles and objects. Later still, students can sort words, that is, the names of the objects.

Venn Diagrams for One-Attribute Sorting This one-circle Venn diagram shows a set of attribute blocks sorted according to whether or not they are squares. The rectangle blocks are outside the circle but inside the rectangle, as they are not squares but are still part of the universal set being sorted.

Shapes that are squares

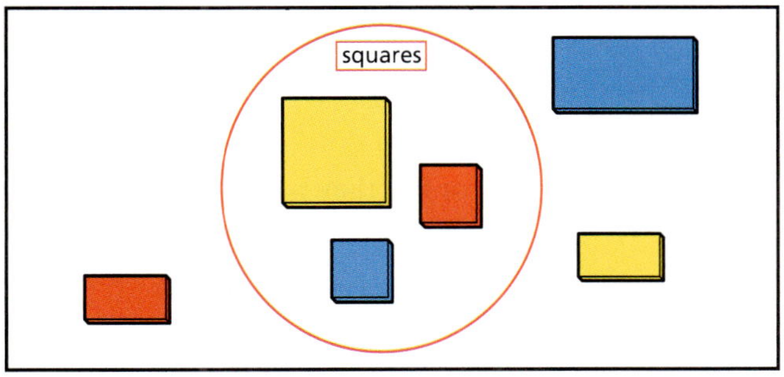

Sorting by one attribute—shape

Venn Diagrams for Two-Attribute Sorting Venn diagrams are excellent tools for sorting by multiple attributes because they make it easy to see which items cross-classify. For example, the buttons in the Venn diagram at the top of the next page have been sorted according to two attributes—colour and number of holes. The buttons in the intersection (where the circles overlap) have both characteristics (two holes and blue). The button outside the circles does not have either characteristic. This sort has two sorting rules, one for each sorting circle.

Teaching Idea 5.5

Point out to students how grocery stores sort food, putting different types of food in different aisles. Have students discuss why this might be useful. Ask why it might make sense to sort a type of food in two different ways and elicit an example or two. [e.g., Raisins might go with fruit (because it is a dried fruit) or it might go with baking items.] Then provide a set of pictures and/or photos of food items for students to sort the way a grocery store might.

Ask students to sort the food cards in one way (e.g., things made of milk vs. things not made of milk). Then have them collect up the cards and sort them in another way (e.g., things you drink vs. things you cannot drink). Finally, have them create a Venn diagram that shows the food items sorted using both characteristics (e.g., things made of milk and things you drink).

To focus on BIOCD 2, ask: *How else could you have sorted the food so there would be more items with both characteristics?* [e.g., things that are fruity and things you drink]

Buttons with two holes and buttons that are blue

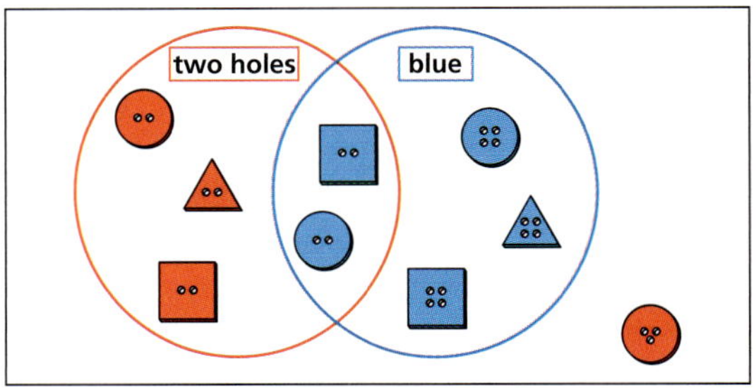

Sorting by two attributes—colour and number of holes

To introduce the notion of how to show cross-classification in Venn diagrams, set out two hoops side by side. Label each hoop with a sorting rule. Make sure the sorting rules and objects to be sorted lend themselves to cross-classification, for example, triangles and square corners (but not, e.g., triangles and quadrilaterals). Have students sort objects into the hoops until they come to an object that cross-classifies. Discuss where the object belongs and elicit the idea of crossing the hoops to form an overlap.

Note that the circles of a Venn diagram do not have to overlap. They can be two discrete circles if the attributes are exclusive, for example, if one hoop contains girls and one contains boys. The circles can also be separate if the items involved in the sort do not exhibit the same characteristics even though they have the potential to do so. For example, if you sort a group of shapes using the characteristics red and striped, the Venn diagram circles remain as separate circles if there are no shapes that are both red and striped. If a striped red shape were included in the set, the circles would have to overlap to accommodate that shape in the intersection of the two circles. It is not wrong to show overlapping circles with nothing in the overlapping part if it is possible that there could be an overlap.

Collecting Data

Asking Good Survey Questions

Young children constantly ask questions to collect data. For example, they might ask their friends whether or not they have a pet dog if they are trying to convince their own parents that they should have a dog since most other kids do. They also watch adults use questions to collect data. For example, a teacher might ask each student, "How many slices of pizza do you want to eat at our party?" so that she knows how much to order for the class.

As you lead discussions about how to conduct surveys to collect data, students will develop skills in collecting useful data. They will begin to see how a well-posed question can provide the data they want, while a poorly posed question might not. For example, if they ask classmates at what time they go to bed, they might encounter difficulties if children go to bed at different times on different nights. If they provide several answer choices, such as "before 7," "between 7 and 8," and "between 8 and 9," they will probably get more manageable data.

Students need to actually experience these problems, rather than simply being told about them, in order to more fully understand the importance of clearly posed questions. Although a teacher might have a tendency to "help" a student by rewriting the questions with them before they conduct the survey, it is probably more helpful for the student if she or he tries to collect data with the poorly posed question and then sees what happens.

Whenever possible, there should be a reason for collecting the data, perhaps in the context of solving a problem, answering a question, or simply to find something out that might be of interest.

Initially, you should encourage students to ask yes/no questions because it is much easier to collect and organize data when only two responses are possible. Even in this simple situation, young students will need to think about how to organize the data they collect.

Later, students can ask questions with two or more possible answer choices, but where the choices are still limited. For example, they might ask classmates to tell whether their favourite season is summer, fall, winter, or spring. Sometimes they will get unexpected answers. For example, if they ask classmates about their favourite pizza toppings, they might not anticipate a choice like anchovies. Show them how to use the category "other" to handle those unexpected answers. Note that, if students get a lot of data in the "other" category, they might need to revise their categories.

Asking questions with multiple possible answers can sometimes lead students to discover that they can get useful data by using questions that involve levels or ratings. For example, they might ask the yes/no question "Can you swim?" Students who respond may have difficulty answering with a simple yes or no, since some are able to swim short distances and others are strong swimmers. The questioner can go back and refine the question by providing levels of swimming ability, including "not at all," as possible answer choices.

Teaching Idea 5.6

Tell students that carnival owners need to think about what rides children like in order to decide which ones to bring to various towns. Ask students to think about how they might collect data about the types of midway carnival rides their classmates like.

To focus on BIOCD 1, ask: *How can you sort rides into different types to help you collect the data?* [e.g., rides that go fast and rides that don't, or rides that spin and rides that don't spin] *How might this information be useful?* [e.g., If you get sick on spinning rides, then you would have lists of all the rides you can and can't go on.]

To focus on BIOCD 3, ask: *How will you ask people a question about rides to make sure it is easy to use the information you get?* [e.g., I will keep a chart with me that has space for tally marks for each of the ride types. I will only let them choose one of the four rides.]

Choosing Data Collection Topics

Primary students are most interested in issues related directly to their own lives. The data they collect should be purposeful. For example, students who are learning about different types of animal life in science might collect and organize data on animals with respect to their colour, size, or habitat.

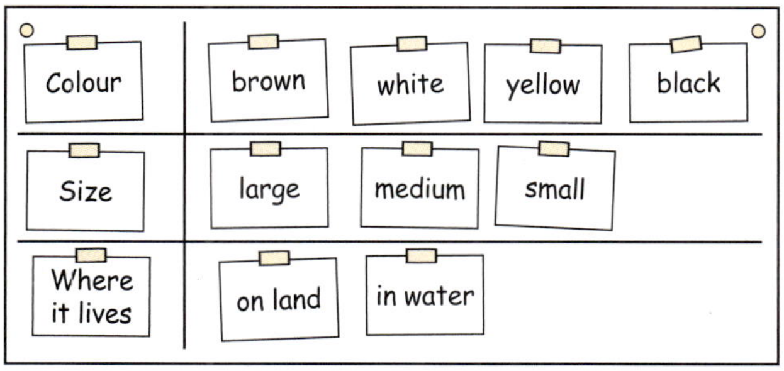

Collecting and organizing data about animals

Here are some ideas for interesting data management topics:

- stories they like to read or hear or their favourite authors
- activities they would like to do in the gym
- extracurricular activities in which they participate
- their favourite cartoons or cartoon characters
- their favourite TV programs or movies
- the number of siblings they have
- the type of pets they have or would like
- how they get to school
- their favourite food (e.g., dessert, breakfast, or ice cream)

Teaching Idea 5.7

Ask students to think about selecting a topic for collecting data from their classmates. Have them tell why they selected the topic they did.

To focus on both BIOCD 1 and BIOCD 3, ask: *How can predicting how your classmates might answer help you figure out how to ask your question?* [e.g., If I ask about the kinds of apples they like, I predict that they will talk about the colour of the apple, so I would ask a question that has apple colour choices like green, yellow, and red.]

Displaying and Analyzing Data

We display data to solve problems or communicate information. A data display, particularly a graph, makes any set of data more accessible and easier to interpret at a glance. Because students are likely to see graphs in the media, a focus on data display helps them become more media literate.

BIG IDEAS FOR DISPLAYING AND ANALYZING DATA

1. Graphs are powerful data displays, since visual displays quickly reveal information about data. (BIDAD 1)
2. Concrete graphs, picture graphs, pictographs, tally charts, bar graphs, and line plots are useful for comparing the frequency of data in different categories. (BIDAD 2)
3. It is important not only to read information from graphs but to make inferences, draw conclusions, and make predictions. (BIDAD 3)

Each teaching idea in this part of the chapter will indicate which Big Idea(s) for Displaying and Analyzing Data (BIDAD) will be emphasized.

Displaying Data

Like data collection, data display involves organizing data so that you can use the data to solve a problem and/or communicate information about a situation.

When students interpret graphs constructed by others, they learn to appreciate the features that can help them make sense of a visual display of data. A good graph communicates some overall impressions of the data at a glance; clear labels and titles and accurately represented data help with this.

Teaching Idea 5.8

Display a concrete graph made of linking cubes and cards marked with the categories "painting" (6 cubes), "playground" (8 cubes), and "sand table" (5 cubes). Add a title card that reads "Our Favourite Activities." Then display the same data in the form of a picture by drawing 6 stick figures near the word "painting," 8 near the word "playground," and 5 near the words "sand table." Make sure the stick figures are scattered within each group so they cannot be subitized. Add the title "Our Favourite Activities."

To focus on BIDAD 1, ask: *Which display would you use to figure out quickly which activity is most popular, the cube graph or the picture? Why?* [The cube graph; I can see which cube train is longer and I do not have to count each cube.]

Data Display Formats

Students need to consider both the audience and the nature of the data when they decide what type of data display to construct. It is important to choose a format that suits the data and to include clear titles and labels to guide the reader who will interpret it. Rather than always telling the student what format to use, it is important to allow students to experiment to see which format works best for them.

The first data displays children usually create are concrete graphs made from real objects. As students begin to recognize that they can use other ways to represent the number of real objects in each category, they progress to graphs with increasing complexity and abstraction—from graphs made with pictures of real objects (picture graphs), to graphs that use symbols (pictographs), and finally to more abstract graphs (bar graphs and line plots).

Concrete Graphs

Using Real Objects Students can use a concrete people graph to answer a simple question about themselves. Notice that students in the concrete graph at the top of the next page are lined up on a floor graphing mat, which is a large square grid. This aligns students so that it is easy to make comparisons. Have your students form people graphs in a way that foreshadows a pictograph or a bar graph with one-to-one correspondence across the categories.

A people graph makes it easy to see that there are 4 more students with long sleeves than with short sleeves.

Other examples of concrete graphs are a graph made with shoes, which compares the number of students wearing running shoes to the number wearing sandals, and a graph made with toy animals to compare the number of students who have cats to the number who have dogs.

Using Representational Objects After students have gained some experience with concrete graphs made with real objects, they are ready to begin making graphs using concrete models that represent the real objects. This adds a level of abstraction because the model is one step removed from the object it represents. For example, to represent the data from the sleeve-length people graph, students could move away from the floor graphing mat, leaving items such as clothespins or linking cubes to hold their places. They could then put the clothespins or cubes together to create a concrete graph. When the students return to their seats, the model will help them recall the data from the people graph.

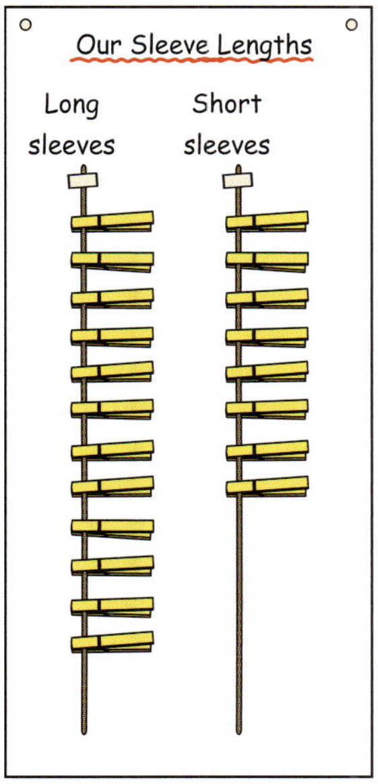

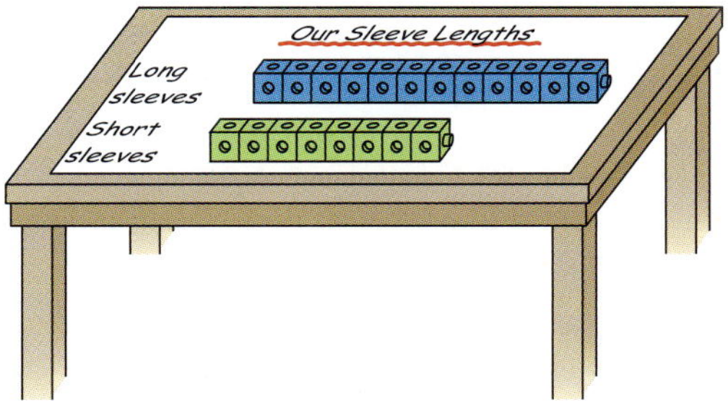

A concrete graph using representational objects is more abstract or symbolic than a people graph.

Teaching Idea 5.9

Provide a pre-made concrete graph made of paper clip chains with no category labels or title. Ask students to describe what the graph might be about. For example, you might place 14 clips in one chain and 10 in the other to show the number of boys and girls in a class.

To focus on BIDAD 1 and 2, ask: *How does this paper clip graph make it easy to compare the two groups?* [I can tell right away which group is bigger without counting.]

Teaching Idea 5.10

Ask students if they would choose to play outside or inside if they were given some "free time" today. Then have them use paper clips to create a graph that shows the results.

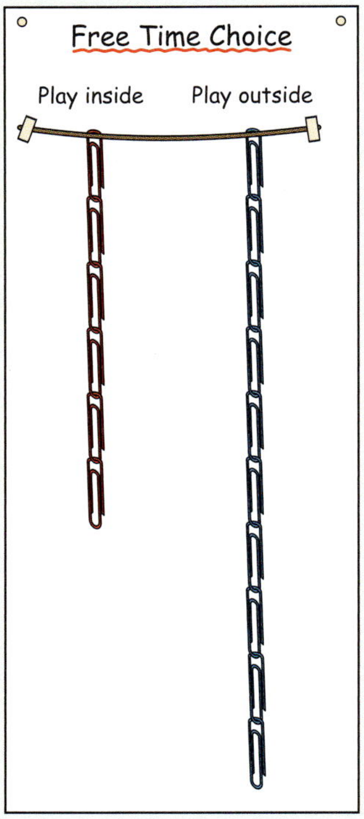

To focus on BIDAD 3, ask: *What does the graph tell you?* [More people want to play outside than inside.]

Encourage students to arrange graph objects in a variety of ways, for example, cubes can be arranged vertically in towers, or horizontally in trains.

As you introduce models of concrete graphs, you should talk about why it helps to make each object the same size or to place the objects on a grid. Doing so ensures that the objects are spaced equally within each row or column and aligned across the categories so you can count and compare them easily. You can provide cubes of different sizes to help make this point. For example, if students were to place different-sized cubes in categories on a concrete representational graph to show whether they have cats, they would discover that a quick visual comparison is impossible.

When you use objects to create a concrete graph and there is no grid, you must align the objects and/or make them uniform.

Useful models or objects to create concrete graphs include

- linking cubes
- counters
- clothes pins
- square tiles
- paper clips
- plastic links

IMPORTANT POINTS ABOUT CONCRETE GRAPHS

- The first person, real object, or concrete model in each category starts at the same level or baseline.
- The people, objects, or models in each category are equally spaced.
- The people or objects in each of the categories are spaced the same way (with or without a grid) or are the same size so that each person or object is aligned with its counterpart in the other categories.
- Concrete graphs can be vertical or horizontal.
- It is important to include labels and a concise but meaningful title to help the reader understand the graph.

Picture Graphs

Before long, students learn that concrete graphs can be unwieldy and difficult to keep intact. They discover that it is more convenient to use pictures to represent data in a picture graph. Picture graphs take graphing to a slightly more abstract level, where students use pictures instead of concrete objects to represent the real thing (each picture is unique to the object or person it represents). Picture graphs form a bridge between concrete graphs and more abstract pictographs and bar graphs. The basic principle of concrete graphs also applies to picture graphs: the reader should be able to make comparisons at a glance.

Picture graphs form a bridge between concrete graphs and more abstract pictographs and bar graphs.

Because the corresponding items in a picture graph must be aligned, you may choose to have students use grid paper to help place the pictures. If the pictures are all the same size, grid paper is not as important. For example, the use of grid paper for the picture graph at the top of the next page on the left is optional.

A WELL-CONSTRUCTED PICTURE GRAPH

When the pieces of data (represented by pictures) in a picture graph are lined up one-to-one across the categories, it is easy to read the graph.

A POORLY CONSTRUCTED PICTURE GRAPH

In this picture graph arrangement, it is hard to compare the categories.

IMPORTANT POINTS ABOUT PICTURE GRAPHS

- Each picture is unique to the individual object or person it represents.
- The first picture in each category starts at the same level or baseline.
- The pictures in each category are equally spaced.
- The pictures in each of the categories are spaced the same way or are the same size so that each picture is aligned with its counterpart in the other categories (grid paper is helpful but optional).
- Picture graphs can be vertical or horizontal.
- It is important to include labels and a concise and meaningful title to help the reader understand the graph.

Pictographs

A pictograph is one step farther than a picture graph along the continuum from concrete to symbolic. We use symbols to represent the data. Sometimes these symbols are picture-like representations. While a picture graph uses a different picture for each piece of data, a pictograph uses the same symbol throughout (although early pictographs use one symbol for each category). The same basic principles of concrete and picture graphs apply: the reader should be able to interpret the graph at a glance by comparing rows or columns.

The following pictograph shows that 10 children were surveyed about their favourite fruits. Six children chose strawberries. The same symbol, a happy face, represents each of the 10 children, although a different colour has been used for each category to reflect the colour of the fruit in that category.

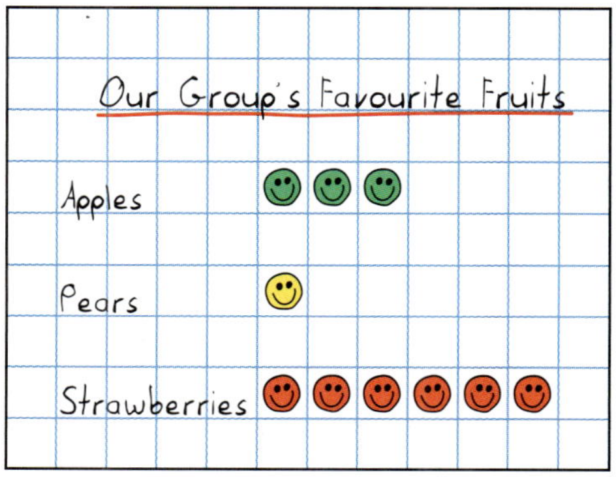

A pictograph uses the same symbol throughout.

Teaching Idea 5.11

Provide cards of several different sizes to students (one card per student). Ask each student to draw a picture of his or her favourite pet on the card. Work with students to create a group picture graph. Talk about how to sort the pictures before graphing them, for example, using the categories "cats," "dogs," and "other."

To focus on BIDAD 1, ask: *Why would it be better if we had all drawn our animals on cards of the same size?* [It would be easier to line up the pictures to see which pet most of us drew.]

A pictograph is one step farther than a picture graph along the continuum from concrete to symbolic.

Teaching Idea | **5.12**

Tell students to think about creating a pictograph that compares the number of students who walk to school, ride the bus, or come by car. Talk about how this information might be useful [e.g., deciding on when to run extracurricular activities].

To focus on BIDAD 1, ask: *Why would a symbol like a school bus not be a good choice to use for the graph?* [It wouldn't be good for the students who walk or come by car, so it would confuse anyone reading the graph.] *What symbol could you use for all three categories? Why?* [e.g., a stick figure, a face, or a picture of a school; Each symbol would represent one student.] *Why is it helpful to line up the symbols?* [I can tell faster which way most students get to school and how many use that way than the other ways.]

Choosing Symbols The graph below uses different symbols in each category. This is not ideal, but it might be expected if students have trouble generalizing to a symbol that pertains to all the categories. One problem with using different symbols is that it is often harder for students to line up the symbols so they are spaced equally.

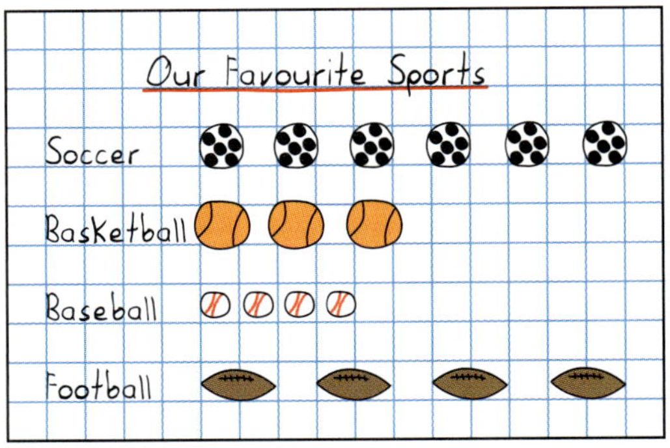

If you use symbols of different sizes, the graph can lose its effectiveness.

Students enjoy choosing symbols for their pictographs, but it can be hard to choose a symbol that represents all the categories. The symbol you choose does not have to reflect the context (e.g., you can use a circle or square), but the graph is more interesting if you choose a meaningful symbol and the symbol provides some immediate information about the data. In the pictograph above, a symbol of a particular type of sport would not be suitable; in fact, it might be misleading. A symbol such as an empty circle would be acceptable, but would not provide as much information as, for example, a happy face. With a happy face, it would be obvious that each symbol represents a person. Another possibility is a T-shirt or a stick figure.

Group discussion and trial and error will help students choose appropriate symbols. You can present some data for students to graph and invite suggestions for possible symbols. Have small groups of students discuss some possibilities and then present their ideas and justifications to the larger group. The class can then talk about why one symbol would be more appropriate than another or why a particular symbol would be inappropriate.

When choosing a symbol for a pictograph, students should consider the ease with which they can draw and copy the symbol. Materials such as stamps, stickers, and stencils can help students create pictographs quickly and efficiently.

IMPORTANT POINTS ABOUT PICTOGRAPHS

- The same symbol is used throughout the graph (although some early pictographs use a different symbol for each category). The symbol may or may not reflect the context of the data.
- The first symbol in each category starts at the same level or baseline.
- The symbols in each category are equally spaced.
- Pictographs can be vertical or horizontal.
- It is important to include labels and a concise but meaningful title to help the reader understand the graph.

Bar Graphs

A bar graph uses the lengths or heights of bars to represent quantities, often the frequencies of particular responses to a data collection activity. Bar graphs are a natural extension of concrete representational graphs made with linking cubes or square tiles, and of pictographs using square pictures and symbols. Students should work on grid paper to draw bar graphs to ensure that all the squares are the same size. To make the transition from concrete graphs to bar graphs drawn on grid paper, you might ask students first to create the bars using square tiles and then to trace around them. Since students are not working with scale at this point, to read the graph many students will count the individual squares in each bar, if they are visible, instead of reading the numerical axis.

Eventually students will read bar graphs where they need to interpret the lengths of the bars, rather than the individual squares that make up the bars, because the grid is missing. For example, to interpret the graph below to compare the number of students with siblings and those without, students can compare the lengths of the bars. However, to determine how many more children have siblings than do not have siblings they must rely on the numerical horizontal axis.

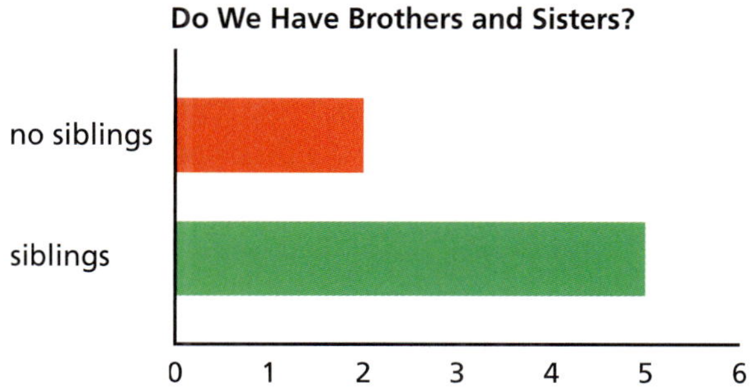

To make the graph easy to read, space the bars an equal distance apart and do not connect them. The separation of the bars also shows that each bar represents a separate or discrete category.

A bar graph can be either horizontal or vertical. Label either the vertical or horizontal axis with the names of the categories and label the other axis with numbers to describe how many are in the category. Although it is not required that you place numbers on the scale axis if each square represents one and the grid is evident, they do make it a bit easier to read the graph. Axis headings for the categories are optional if it is obvious from the category labels and title what the graph is about. In the graph below, the axis heading "Dinosaur" could be considered optional because of the graph's title.

HORIZONTAL BAR GRAPH

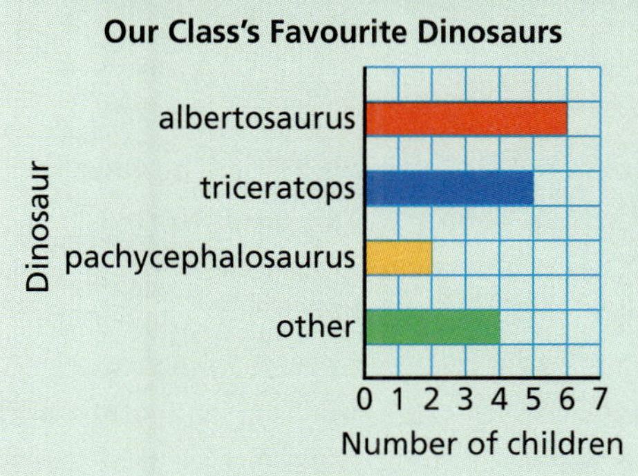

VERTICAL BAR GRAPH

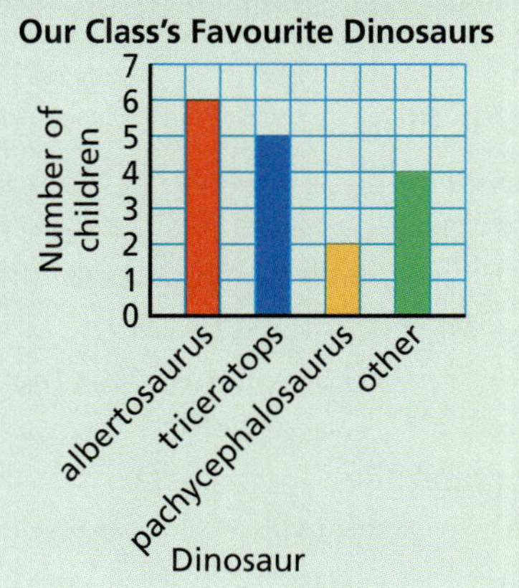

Teaching Idea 5.13

Provide data or have students collect data that they can use to compare subgroups of students in the class. For example, they might compare the number of students whose first names have fewer than 6 letters, 6 to 8 letters, and 8 or more letters to decide how big name cards should be. They can create a bar graph to show the data they have collected.

To focus on BIDAD 1, ask: *Why is it important to make the bars out of squares that are the same size?* [Otherwise, it would be hard to look to see which group is bigger.]

IMPORTANT POINTS ABOUT BAR GRAPHS

- Bars should be separated to indicate that they represent discrete data, although it is not incorrect if there are no spaces between the bars.
- If there are no pieces of data for a category or interval, the category can be dropped entirely from the graph or you can include the category name but leave a space where the bar would be, depending on the context.
- Bar graphs can be vertical or horizontal.
- Each bar should have a category label, which might be a discrete topic name or number. The other axis, the scale axis, can be labelled numerically but this is optional if the graph is on a grid and the scale is 1:1.
- Axis headings should be used, as necessary, for clarity.
- It is important to include a concise but meaningful title to help the reader understand the graph.

Teaching Idea 5.14

Prepare a tally chart like the one below, where the tallies are not lined up. Tell students this was the result of asking 26 friends whether or not they like mangos to decide whether to have mangos at a class party.

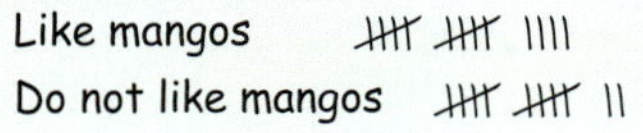

To focus on BIDAD 2, ask: *How does the tally chart show how many people like and do not like mangos?* [You count the number of tally marks in each category.]

To focus on BIDAD 3, ask: *How does the tally chart show how many more like mangos than do not like mangos?* [You count how many more tallies are in the first row than in the second row.]

To focus on BIDAD 1, ask: *Why would lining up the tally marks like this make the chart easier to use?*

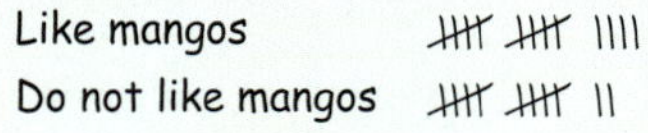

[You could see by just looking that 2 more people like mangos without having to count every tally in each row.]

Tally Charts

A tally chart is a fairly simple, but somewhat abstract, way to display data. Students use tally marks to count the frequency of responses or items in different categories. Grouping the tally marks in 5s makes it easier for students to skip count the number in each category. Notice how the tallies below are aligned in columns for easy comparison. From this tally chart, it is easy to see at a glance that more people chose the first story than the second story. Every tally chart needs a title that clearly explains the meaning of the data.

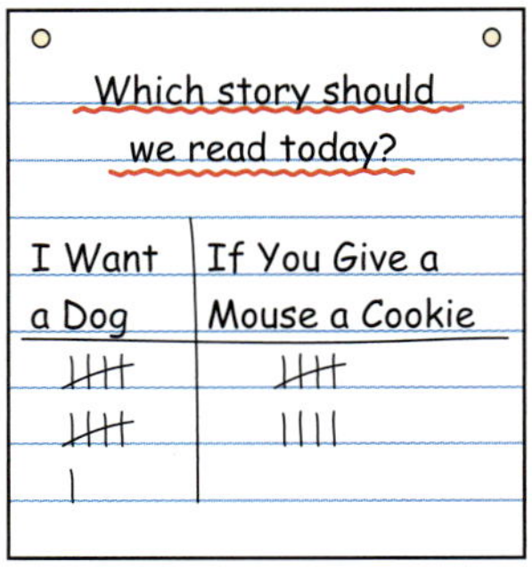

When you line up the tally marks like this, it is easy to tell at a glance which book is more popular.

IMPORTANT POINTS ABOUT TALLY CHARTS

- Different categories should be represented in parallel rows or columns.
- It is useful to use separate rows (or columns) for each group of 5 tallies, if possible.
- Tally charts can be vertical or horizontal.
- It is important to include labels and a concise but meaningful title to help the reader understand the chart.

Line Plots

A line plot shows the frequency of data by organizing it along a number line. We usually use an X symbol to represent each piece of data. Normally the number line is horizontal and the Xs are in vertical columns.

A line plot can serve as both the data display and the data collection and organization tool.

To collect the data for the line plot at the top of the next page, 22 people were asked how many visits they made to a convenience store last week. A line plot can serve as both the data display and the data collection and organization tool; it is an expedient way to manage data. Notice that this is like a pictograph with one-to-one correspondence; each X is like a symbol. Because a line plot is based on a number line, you can only use a line plot in a situation where the categories are numbers.

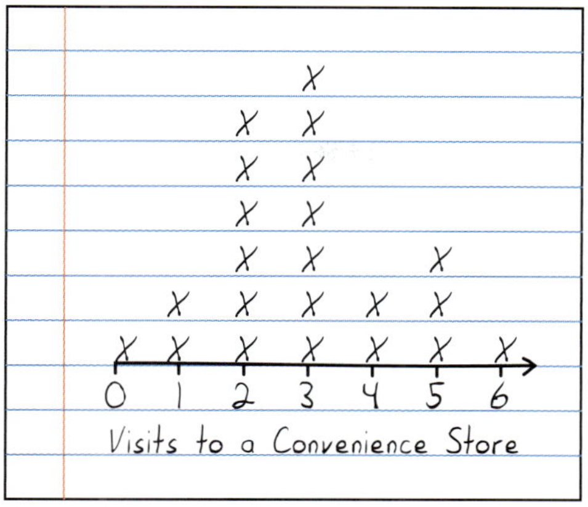

A line plot is a graph that uses a number line to organize numerical categories.

Each X on the line plot above represents one person. It appears that most of the people visited a convenience store 2 or 3 times, and only a few people went a different number of times. The graph shows this pattern visually.

To help students recognize when line plots are appropriate or useful, you can provide a number of different scenarios and ask students to decide whether or not it makes sense to use a line plot to display the numbers in those situations. For example, you might ask if they would use a line plot to show: how many books they read in the past week or month (yes); whether they prefer chapter books or picture books (no); how many pages were in the last book they read (yes, although the line might be long, or it might skip count by 5s or 10s rather than showing every number).

IMPORTANT POINTS ABOUT LINE PLOTS

- The categories for which data is being collected must be numerical.
- The size of the number line required must be reasonable (e.g., the categories might be 1 to 10, but not numbers like 1, 2, 100, and 1000).
- The Xs above the numerical categories should be lined up (lined or grid paper might be helpful).
- The plots could be horizontal, although they are conventionally vertical.
- It is important to include labels and a concise but meaningful title to help the reader understand the chart.

Teaching Idea 5.15

Students might create a line plot to show the number of siblings of each student in the class.

To focus on BIDAD 2, ask: *Why does it make sense to use a line plot to show this set of data?* [e.g., The categories are numbers of siblings that go 0, 1, 2, 3 so I can put them along a number line; The line plot is good for collecting data values and showing them in a graph at the same time.] *Could you use another type of graph instead? Explain.* [Yes; I could use a pictograph or a bar graph. Each category would be a number.]

Analyzing Data

Interpreting Graphs

When students interpret graphs constructed by others, they learn to appreciate the features that help them make sense of a visual display of data. A good graph should communicate at a glance some overall impressions of the data. To do this, you must choose a graph type that suits the data, provide clear labels and titles, and represent the data accurately. The following sample graphs represent different levels of complexity. Each is constructed to make it easy to interpret the data.

A good graph should communicate at a glance some overall impressions of the data.

Concrete Graphs

The concrete graph at the top of the next page allows the reader to see quickly that there are more green cubes than red or black cubes because it uses cubes of the same size in all categories, starts each row of cubes on a common baseline, and has clear labels and a title.

Teaching Idea 5.16

Provide the same data to students in two different graphs (concrete graph, picture graph, pictograph, or bar graph), one that is constructed horizontally and one that is constructed vertically.

To focus on BIDAD 3, ask: *Do the graphs tell me the same information or different information? How do you know?* [The same information; e.g., I read how many people were in each category and it was the same for both.]

Is it always possible to use a horizontal graph to show the same data from a vertical graph? [Yes, if your paper is wide enough.]

Teaching Idea 5.17

Provide an assortment of bar graphs to groups of students. Ask them to tell the number in each category of each graph they receive. Observe whether students always count from the baseline of the graph or whether they use information from one category to help them to read information in another category.

To focus on BIDAD 3, provide a bar graph like the one shown on the next page and ask: *How do you know that 6 people prefer chocolate chips as toppings? Is there more than one way to find out? Explain.* [e.g., I counted the number of squares in the bar labelled "chocolate chips"; yes, I could read across from the number line at 6.] *How can you use more than one way to figure out how many chose "berries"?* [e.g., I could read across from the number line at 4, or I could count down 2 from 6, or I could count the 4 squares in the red bar.]

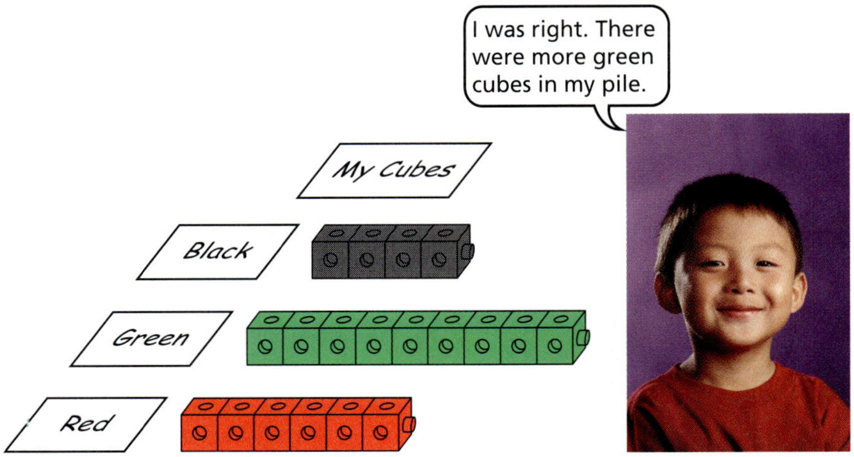

A well-constructed concrete graph makes it easy to compare categories.

Picture Graphs

This picture graph is easy to interpret because it uses a grid to make sure the pictures in each category are equally spaced, each column of pictures starts on a common baseline, and there are clear labels and a title.

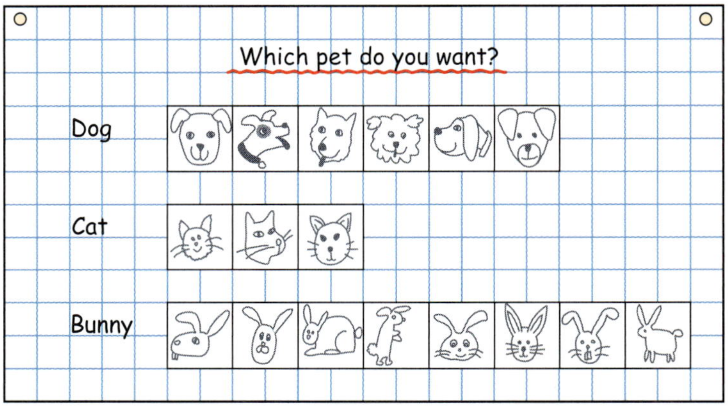

A well-constructed picture graph makes it easy to extract information.

Pictographs

A reader can easily interpret the symbols in this pictograph because it uses the same symbol throughout, the symbols in each row start on a common baseline, the symbols in the rows are aligned, and it has clear labels and a title.

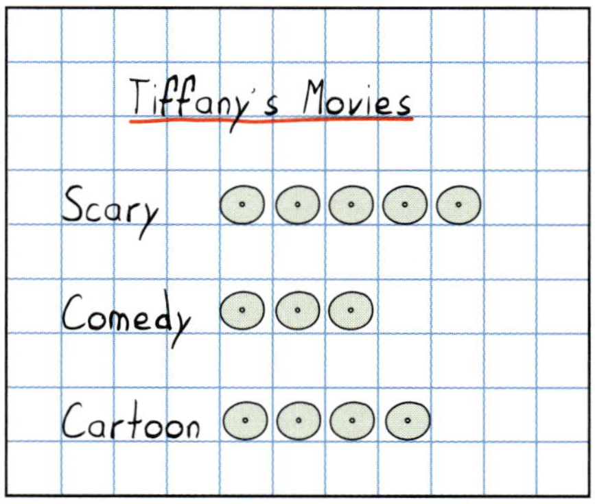

Bar Graphs

This bar graph allows the reader to interpret the data easily because it is drawn on a grid, each bar starts on a common baseline, and it has clear labels and a title. This graph uses a different colour for each bar. Note that this is optional and is often done simply to make the graph more attractive (although sometimes the colour chosen for each bar reflects the context of the category in some way).

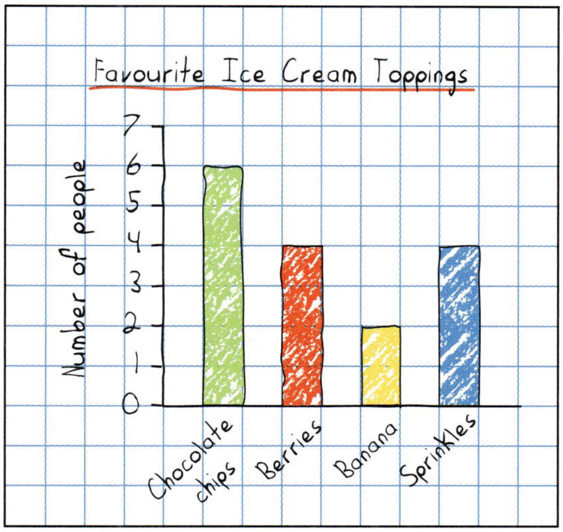

Reading, Drawing Conclusions, and Making Inferences from Graphs

One of the reasons teachers use visual displays of data is to paint a picture that reveals more about the data than would the numerical data alone. Your goal is for your students to be able to read individual specific pieces of information or facts from a graph and also to work with multiple facts at the same time, for example, to make comparisons, and draw conclusions. Eventually, you want them to make inferences and predictions about what they see on the graph, using appropriate justification.

The types of conclusions students might draw often involve comparisons (e.g., which category has two more than which other category or which category has more than double the number of which other category) or combinations (e.g., how many chose A or B). The types of inferences and predictions students might make could involve predicting whether the same results are likely in another group or predicting what is likely to happen if more data is added to the existing information.

The example below is based on the ice cream topping graph above. It shows the sort of information you can glean from graphs:

READING THE GRAPH	CONCLUSIONS FROM THE GRAPH
• 6 people like chocolate chips • 4 people like berries • 2 people like bananas • 4 people like sprinkles	• 2 more people like chocolate chips than like berries • 4 fewer people like banana than like chocolate chips • 16 people answered the question about favourite ice cream toppings (6 + 4 + 2 + 4 = 16)

Possible inferences, or predictions, based on this set of data might be

- In a Grade 6 class, I think that students will still pick chocolate chips the most, since I know that my brother's friends like chocolate chips too.
- If we asked five more Grade 2 students, I think two or three of them would also pick chocolate chips.

Teaching Idea 5.18

For an interesting activity, provide two bar graphs without titles or labels and ask students to infer what they might be about. For example,

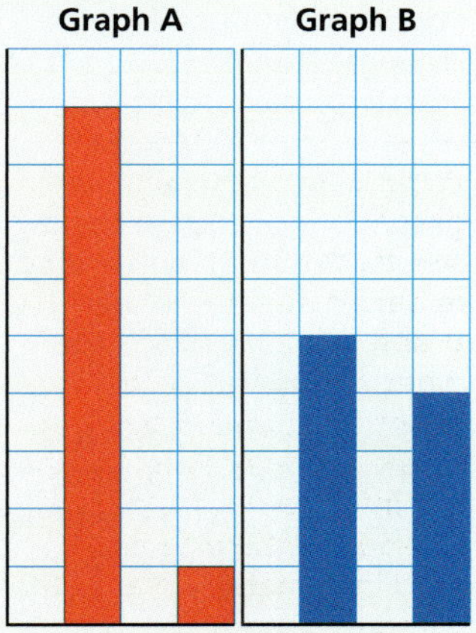

To focus on BIDAD 3, ask: *Which graph compares the number of adults to children on a team? Which graph compares the number of 7-year-olds to 6-year-olds on that team? Why do you think that?* [Graph A, since there are lots of children and only one coach on a team; Graph B, since there are about the same number of 6-year-olds as 7-year–olds.]

To focus on BIDAD 1, ask: *How does the first graph make it obvious that a team has lots more children than adults?* [The bars are so different in size that I noticed right away.]

Probability

Probability is the study of uncertainty. Mathematicians apply mathematical principles to describe and predict in situations involving uncertainty. Even young students can make sense of some simple probability concepts.

Each teaching idea in this part of the chapter indicates which Big Idea(s) for Probability (BIPr) you can emphasize.

BIG IDEAS FOR PROBABILITY

1. Categorizing events using terms such as *likely, certain, impossible*, and *unlikely* is a way of describing probability. (BIPr 1)
2. In experimental situations, we base our predictions on past events. (BIPr 2)
3. In a probability situation, you can never be sure what will happen next. (BIPr 3)

Early Work with Probability

Young students do not usually discuss theoretical probability, for example, the notion that the probability of flipping a coin and getting heads is $\frac{1}{2}$ or that it is as likely to flip heads as to flip tails. Instead, they use simple language to describe things that are certain, impossible, or somewhere in between (likely or unlikely). As they conduct experiments by spinning coloured spinners, drawing coloured cubes out of a bag, or rolling a single die, they begin to get a sense of what is likely to happen. They may even do some simple predicting based on the results of an experiment.

Initially, you should ask students to think about familiar events in their lives and describe them using vague, informal probability words. This begins probability work at a concrete level. In this sense, concrete means familiar contexts that are relevant to students. You can have them describe the probability of events like these:

- the sun will rise tomorrow ("for sure")
- a parent will be home when you get home from school ("probably")
- you will have spaghetti for dinner ("maybe")
- an elephant will come to school on Friday ("impossible!")

You can model and ask young students to describe events like these using probability language such as *always happens* and *never happens* as well as *likely to happen* and *unlikely to happen.*

You have some flexibility in the probability language you choose to introduce. While some teachers use the term *always*, others use *certain.* Some teachers use the term *impossible*, while others use *never.*

Students are sometimes confused by the incongruity between the way particular words are used in everyday language and the way these terms are used in mathematics. In mathematics, an event is called *likely* if its probability is greater than $\frac{1}{2}$. In everyday situations, we tend to reserve the word *likely* for events that are very likely to happen.

A display like the one at the top of the next page can help students keep track of and organize new vocabulary. You might write the terms on separate cards on a word wall to allow for some flexibility in how they are organized.

Teaching Idea 5.19

Ask students to sort themselves using probability ideas to create a people graph. For example, one category could be students who are certain to play sports this week (Group A), another could be students who are likely but not certain to play sports (Group B), and a third could be those who are unlikely to play sports this week (Group C).

To focus on BIPr 1, ask: *Is it likely that each student in Group A will spend the same amount of time playing sports this week? Explain your thinking.* [I can't tell; The graph shows that they will play sports not how long they will play.]

To focus on BIPr 2, ask: *Suppose we made this same graph in the other Grade 2 class. Do you think the results would be similar? Explain your thinking.* [I think so; They are all kids the same age.]

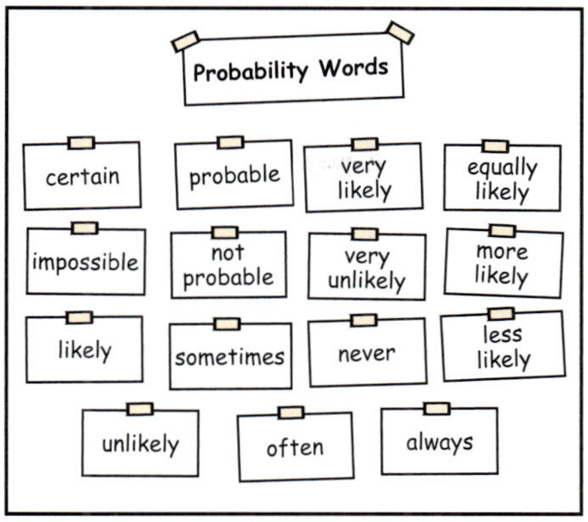

A probability word wall

Experimental Probability

Students are often curious and interested in things that cannot be predicted with certainty, like how a spinner will land. They experience these situations in the context of games in their everyday lives. Exploration of these probabilities is motivating and relevant to them.

Most of the probability studied at the elementary level involves experiments; these concrete activities are at a more appropriate developmental level for these young students. The transition to theoretical probability comes later when students are better able to deal with more abstract notions.

It makes sense to have students make predictions and then test their predictions when conducting experiments. For example, they might predict and test the following:

- When they spin the spinner, which colour are they more likely to land on?
- When they roll a die, does one number come up more often than another?
- When they roll two dice, does each sum come up equally often?
- When they draw from a bag of coloured cubes multiple times, are they more likely to draw one colour than another?

Even in the earliest grades, students can conduct simple experiments. For young students, however, you will need to define exactly what they must do.

To conduct typical experiments, students need to use probability devices that produce two or more randomly occurring outcomes—outcomes that cannot be predicted with any certainty.

RANDOM DEVICES INCLUDE ...	TYPICAL EXPERIMENTS MIGHT INVOLVE ...
• paper cups	• tossing a paper cup repeatedly to see how often it lands on its side
• coins or 2-sided tiles	• flipping a coin or 2-sided tile repeatedly and counting the number of times each face comes up
• coloured cubes and bags	• drawing coloured cubes repeatedly from an opaque bag and keeping track of the colours drawn
• dice	• rolling a die repeatedly and keeping track of the numbers rolled
• spinners	• repeatedly spinning spinners with coloured or numbered sectors, or sectors with different pictures on them, and keeping track of the results

Teaching Idea 5.20

Have students predict what colour they will spin most often on the spinner below and explain their prediction. Then have them spin 20 times and record the resulting colours by selecting linking cubes of the same colours and then sorting them. Ask them which colour they spun more often, yellow or blue, and how they know.

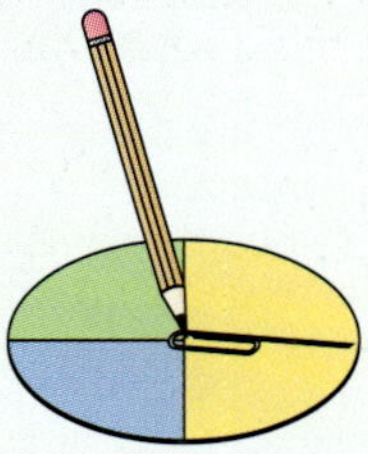

To focus on BIPr 2, ask: *Suppose I spin 5 more times. Do you think I will spin more yellows or more blues? Why?* [More yellows; I got more yellows the first time, or the yellow section is bigger on the spinner.]

To focus on BIPr 3, ask: *Last time you spun 3 yellows in 5 spins. Can you be sure that will happen again?* [No, I can't be sure since each time different things can happen.]

As discussed in the earlier section about data collection, students need to attend to how to keep track of the data they collect and record it for later examination.

Collecting and Organizing Probability Data

At the K to 3 level, it is helpful to provide organizational tools for students to use to keep track of the results of their experiments. For example,

- If they are using spinners with coloured sections and keeping track of colours spun in, say, 10 spins, you might provide linking cubes or counters of those colours that they can put in front of them to show the colour they spun each time. After 10 spins, they can sort the cubes by colour to see what happened most often, least often, and so on.
- If students are recording numbers rolled on a die or numbers or letters spun on a spinner, you might provide a simple chart like this for them to record their results.

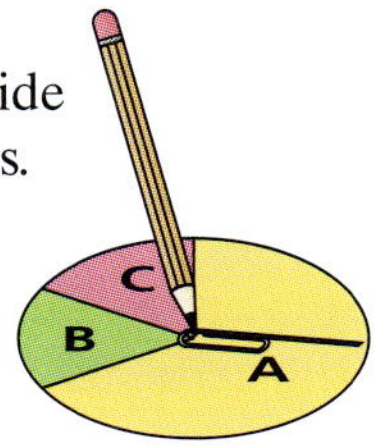

Spin number	1	2	3	4	5	6	7	8	9	10
Letter I spun	A	A								

It is also important to suggest or, even better, allow students to suggest or develop different ways to keep track of the collected data. Some possible approaches are shown below.

Sample Probability Experiments

If students are tossing a paper cup repeatedly to see if it lands on its side most of the time, they could record their results by creating a picture graph.

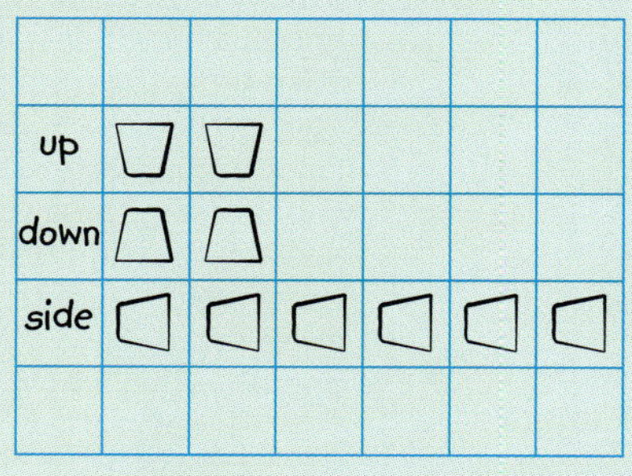

If students are flipping a coin or a 2-sided tile repeatedly to see which face comes up more often, they could record their results in a tally chart.

If students are drawing coloured cubes repeatedly from an opaque bag to figure out what colour cubes are in the bag, they could make a permanent record of their results by creating a concrete cube graph of the results using cubes of the same colour.

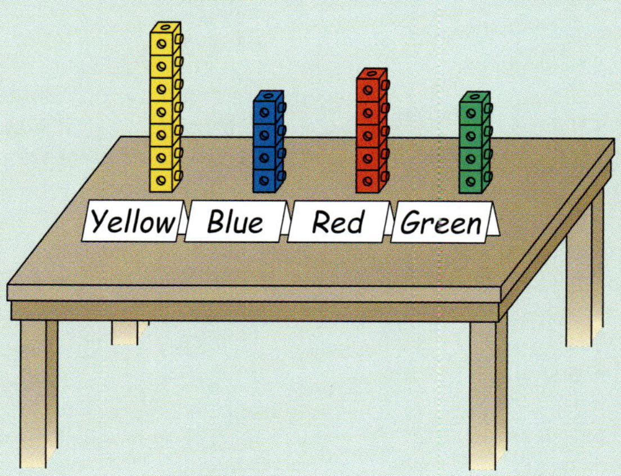

If students are rolling a die repeatedly to see if any number is more likely than another, they could record their results by using a line plot from 1 to 6.

Making Predictions Based on Experimental Results

Making predictions is an important part of any experiment. You want students to use prior experiences to help them predict what will happen next. To introduce the notion that students should not make predictions before doing multiple experiments or trials, encourage them to use at least 10 or more trials of an experiment. Discuss informally why it might be harder to predict what is likely to happen after 1 or 2 trials than after, for example, 10 trials.

Note that young children sometimes intuitively use theoretical probability to predict in very simple situations. For example, if they are spinning a spinner that is $\frac{7}{8}$ yellow and $\frac{1}{8}$ red, they will predict that the needle will land on yellow more often because "there is more yellow than red on the spinner." This is theoretical probability because it results from analyzing the situation rather than from conducting an experiment. However, the focus should be on conducting an experiment to test their prediction. They may be surprised to discover that it lands on red at all.

Students need to learn that it is all right if their predictions do not come true; in fact, it is critical to their understanding of the concept of probability for them to see that they can never be certain of what will happen. Still, students should be able to explain or justify the predictions they make.

It is critical to students' understanding of the concept of probability for them to see that they can never be certain of what will happen.

References

Bamberger, H., and Hughes, P. (1999). *Super graphs, Venns and glyphs.* New York: Scholastic.

Baroody, A.J. (1992a). The development of preschoolers' counting skills and principles. In Bideaud, J., Meljac, C., and Fischer, J.P. (Eds.). *Pathways to number: Children's developing numerical abilities.* Hillsdale, NJ: Erlbaum, 99–126.

Baroody, A.J. (1992b). Remedying common counting difficulties. In Bideaud, J., Meljac, C., and Fischer, J.P. (Eds.). *Pathways to number: Children's developing numerical abilities.* Hillsdale, NJ: Erlbaum, 307–324.

Baroody, A.J., and Benson, A. (2001). Early number instruction. *Teaching Children Mathematics, 8,* 154–158.

Baroody, A.J., and Dowker, A. (2003). *The development of arithmetic concepts and skills.* Mahwah, NJ: Lawrence Erlbaum Associates.

Baroody, A.J., and Wilkins, J.L.M. (1999). The development of informal counting, number, and arithmetic skills and concepts. In Copley, J. (Ed.). *Mathematics in the early years.* Reston, VA: National Council of Teachers of Mathematics, 48–65.

Bialystok, E., and Codd, J. (1996). Developing representations of quantity. *Canadian Journal of Behavioural Science, 28,* 281–291.

Bogart, J.E. (1989). *Ten for dinner.* Richmond Hill, ON: Scholastic Tab.

Borko, H., and Putnam, R. (1995). Expanding a teacher's knowledge base: Cognitive psychological perspective on professional development. In Guskey, T., and Huberman, M. (Eds). *Professional development in education: New paradigms and practices.* New York: Teacher's College Press, 35–65.

Buchholz, L. (2004). Learning strategies for addition and subtraction facts: The road to fluency and the license to think. *Teaching Children Mathematics, 10,* 362–367.

Burns, M. (1994). *The greedy triangle.* New York: Scholastic Press.

Carpenter, T.P., Fennema, E., Franke, M.L., Levi, L., and Empson, S. (1999). *Children's mathematics: Cognitively guided instruction.* Portsmouth, NH: Heinemann.

Cavanagh, M., Dacey, L., Findell, C.R., Greenes, C.E., Sheffield, L.J., and Small, M. (2004). *Navigating through number and operations in Prekindergarten–grade 2.* Reston, VA: National Council of Teachers of Mathematics.

Charles, R. (2004). Big ideas and understandings as the foundation for elementary and middle school mathematics. *Journal of mathematics education leadership, 7,* 9–24.

Christelow, E. (1998). *Five little monkeys jumping on the bed.* New York: Clarion Books.

Clarke, D., Cheeseman, J., McDonough, A., and Clark, B. (2003). Assessing and developing measurement with young children. In Clement, D., and Bright, G. (Eds.). *Learning and teaching measurement.* Reston, VA: National Council of Teachers of Mathematics, 68–80.

Clements, D.H. (1999). Geometric and spatial thinking in young children. In Copley, J. (Ed.). *Mathematics in the early years.* Washington, DC: National Council of Teachers of Mathematics and National Association for the Education of Young Children, 66–79.

Clements, D.H. (1999). Subitizing: What is it? Why teach it? *Teaching Children Mathematics, 5*, 400–405.

Clements, D.H., and Sarama, J. (2000a). The earliest geometry. *Teaching Children Mathematics, 7*, 82–86.

Clements, D.H., and Sarama, J. (2000b). Young children's ideas about geometric shapes. *Teaching Children Mathematics, 6*, 482–487.

Clements, D.H., and Sarama, J. (Eds.), DiBiase, A-M. (Assoc Ed.) (2004). *Engaging young children in mathematics: Standards for early childhood mathematics education*. Mahwah, NJ: Lawrence Erlbaum Associates.

Cook, C.D. (2000). I scream, you scream: Data analysis with kindergartners. *Teaching children mathematics, 14*, 538–540.

Dacey, L., Cavanagh, M., Findell, C., Greenes, C., Sheffield, L.J., and Small, M. (2003). *Navigating through measurement, prekindergarten–grade 2*. Reston, VA: National Council of Teachers of Mathematics.

Dark, D. (2007-2008). Making "cute" count. *Teaching children mathematics, 14*, 153–159.

Findell, C.R., Small, M., Cavanagh, M., Dacey, L., Greenes, C.E., and Sheffield, L.J. (2001). *Navigating through geometry in prekindergarten–grade 2*. Reston, VA: National Council of Teachers of Mathematics.

Flowers, J., Kline, K., and Rubenstein, R.N. (2003). Developing teachers' computational fluency: Examples in subtraction. *Teaching Children Mathematics, 9*, 330–334.

Fosnot, C.T., and Dolk, M. (2001). *Young mathematicians at work: Constructing number sense, addition, and subtraction*. Portsmouth, NH: Heinemann.

Frye, D., Braisby, N., Lowe, J., Maroudas, C., and Nicholls, J. (1989). *Young children's understanding of counting and cardinality*. Child Development, *60*, 1158–1171.

Fuson, K.C. (1988). Children's counting and concepts of number. New York: Springer-Verlag.

Fuson, K.C. (1990a). Conceptual structures for multiunit numbers: Implications for learning and teaching multi-digit addition, subtraction, and place value. *Cognition and Instruction*, 7, 343–403.

Fuson, K.C. (1990b). Issues in place-value and multi-digit addition and subtraction learning and teaching. *Journal for Research in Mathematics Education, 21*, 273–280.

Gelman, R., and Gallistel, C.R. (1978). *The child's understanding of number*. Cambridge, MA: Harvard University Press.

Goodrow, A.M., and Kidd, K. (2008). "We all have something that has to do with tens": Counting school days, decomposing number, and determining place value. *Teaching Children Mathematics, 15*, 74–79.

Greenes, C., and Findell, C. (1999b). *Groundworks: Algebraic thinking, grades 1–3*. Chicago: Creative Publications.

Greenes, C., Cavanagh, M., Dacey, L., Findell, C., and Small, M. (2001). *Navigating through algebra in pre-kindergarten–grade 2*. Reston, VA: National Council of Teachers of Mathematics.

Hannibal, M.A. (1999). Young children's developing understanding of geometric shapes. *Teaching Children Mathematics, 5*, 353–357.

Harris, T. (2000). *Pattern fish*. Brookfield, CN: Millbrook Press.

Harris, T. (2001). *Pattern bugs*. Brookfield, CN: Millbrook Press.

Hewitt, S. (2000). *Pirates at sea*. London: Belitha Press Ltd.

Hiebert, J. (1984). Children's mathematics learning: The struggle to link form and understanding. *Elementary School Journal, 84*, 496–513.

Hoban, T. (1998). *So many circles, so many squares*. New York: Greenwillow Books.

Huber, L.L., and Lenhoff, R.S. (2006). Mathematical concepts come alive in pre-k and kindergarten classrooms. *Teaching children mathematics, 13,* 226–231.

Huinker, D., Freckman, J.L., and Steinmeyer, M.B. (2003). Subtraction strategies from children's thinking: Moving toward fluency with greater numbers. *Teaching Children Mathematics, 9,* 347–353.

Kalifatidou, E.R. (2007-2008). Design of activities on numerical representations based on cognitive research. *Teaching children mathematics, 14,* 355–360.

Kamii, C. (1986). Place value: An explanation of its difficulty and educational implications for the primary grades. *Journal of Research in Childhood Education, 1,* 75–86.

Kamii, C., and Dominick, A. (1999). The harmful effects of algorithms in grades 1–4. In Morrow, L., and Kenney, M.J. (Eds.). *The teaching and learning of algorithms in school mathematics.* Reston, VA: National Council of Teachers of Mathematics, 130–140.

Kamii, C., Lewis, B.A., and Booker, B.M. (1998). Instead of teaching missing addends. *Teaching Children Mathematics, 4,* 458–461.

Kamii, C., Lewis, B.A., and Livingston, S.J. (1993). Primary arithmetic: Children inventing their own procedures. *Arithmetic Teacher, 41,* 200–203.

Kribs-Zaleta, C.M., and Bradshaw, D.L. (2003). A case of units. *Teaching Children Mathematics, 9,* 397–399.

Leutzinger, L.P. (1999). Developing thinking strategies for addition facts. *Teaching Children Mathematics, 6,* 14–18.

McMillan, B. (1991). *Eating fractions.* New York: Scholastic Press.

Murphy, S. (2001). *Captain Invincible and the space shapes.* New York: Harper Trophy.

Myller, R. (1991). *How big is a foot?* New York: Yearling.

Novakowski, J. (2007). Developing "five-ness in kindergarten." *Teaching children mathematics, 14,* 226–231.

Postlewait, K.B., Adams, M.R., and Shih, J.C. (2003). Promoting meaningful mastery of addition and subtraction. *Teaching Children Mathematics, 9,* 354–357.

Sarama, J., and Clements, D.H. (2003). Building blocks of early childhood mathematics. *Teaching Children Mathematics, 9,* 480–484

Schifter, D., Bastable, V., and Russell, S.I. (1997). Attention to mathematical thinking: Teaching to the big ideas. In Friel, S., and Bright, G. (Eds.). *Reflecting on our work: NSF teacher enhancement in mathematics K–6.* Washington, DC: University Press of America, 255–261.

Schwerdtfeger, J.K., and Chan, A. (2006). Counting collections. *Teaching children mathematics, 13,* 356–361.

Sheffield, L.J., Cavanagh, M., Dacey, L., Findell, C., Greenes, C., and Small, M. (2002). *Navigating through data analysis and probability, prekindergarten–grade 2.* Reston, VA: National Council of Teachers of Mathematics.

Small, M. (2005a). *PRIME: Number and operations: Background and strategies.* Toronto: Thomson Nelson.

Small, M. (2005b). *PRIME: Patterns and algebra: Background and strategies.* Toronto: Thomson Nelson.

Small, M. (2006). *PRIME: Data management and probability: Background and strategies.* Toronto: Thomson Nelson.

Small, M. (2007). *PRIME: Geometry: Background and strategies.* Toronto: Thomson Nelson.

Small, M. (2010). *PRIME: Measurement: Background and strategies.* Toronto: Nelson Education Ltd.

Sophian, C. (2002). Learning about what fits: Preschool children's reasoning about effects of object size. *Journal of Research in Mathematics Education, 33,* 290–302.

Steffe, L.P. (1992). Learning stages in the construction of the number sequence. In Bideaud, J., Meljac, C., and Fischer, J.P. (Eds.). *Pathways to number: Children's developing numerical abilities.* Hillsdale, NJ: Lawrence Erlbaum, 83–98.

Stephan, M., and Clements, D.H. (2003). Linear and area measurement in prekindergarten to grade 2. In Clements, D.H., and Bright, G. (Eds.). *Learning and teaching measurement.* Reston, VA: National Council of Teachers of Mathematics, 3–16.

Tank, B. (1995). *Math by all means: Probability, grades 1–2.* Sausalito, CA: Marilyn Burns Education Associates.

Tompert, A. (1990). *Grandfather Tang's story.* New York: Crown Publishers.

Usiskin, Z. (1997). *Doing algebra in grades K–4.* Teaching Children Mathematics, 3, 346–356.

Whitenack, J.W., Knipping, N., Novinger, S., and Underwood, G. (2001). Second graders circumvent addition and subtraction difficulties. *Teaching Children Mathematics, 8,* 228–233.

Whitin, P., and Whitin, D.J. (2008). Learning to solve problems in primary grades. *Teaching children mathematics, 14,* 426–432.

Wright, T., Mokros, J., and Russell, S. J. (1997). *Measuring: Bigger, taller, heavier, smaller; Grade level 1; Investigations in number, data, and space.* Palo Alto, CA: Dale Seymour Publications.

Yackel, E. (1997). A foundation for algebraic reasoning in the early grades. *Teaching Children Mathematics, 3,* 276–280.

GLOSSARY

2-D shape: a geometric figure that has two dimensions; e.g., a triangle or a circle

3-D shape: a geometric figure that has three dimensions; e.g., a prism or a sphere; sometimes called an object, a figure, or a solid

addend: one of the numbers added in an addition equation; e.g., in $3 + 4 = 5$, the addends are 3 and 4

algorithm: a systematic procedure or set of steps, usually for carrying out a computation; there are multiple algorithms for any operation

apex: the highest point of a geometric figure relative to a base; an apex can be a vertex or not; e.g., the apex of a pyramid is also one of its vertices, but the apex of a cone is not a vertex

area: the number of uniform two-dimensional units that will cover a 2-D shape

arithmetic sequence: an increasing (growing) or decreasing (shrinking) number pattern in which the difference between consecutive terms is constant; e.g., for 3, 8, 13, 18, 23, ..., the constant difference is 5

associative property: a property of both addition and multiplication whereby you can change the grouping of the numbers without changing the result, i.e., $(a + b) + c = a + (b + c)$ and $(a \times b) \times c = a \times (b \times c)$

attribute: a characteristic or quality of a pattern or shape, such as colour, size, mass, and number of vertices. See *geometric attribute* and *measurement attribute.*

bar graph: a visual way to display data that uses horizontal or vertical bars to represent data values grouped in categories; e.g.,

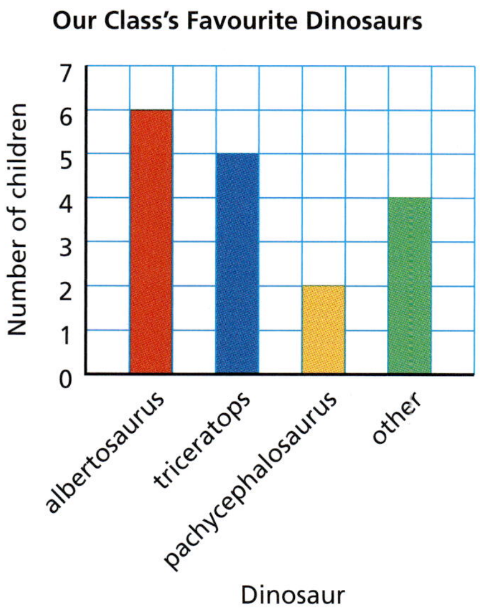

base: **1.** for a triangle or parallelogram, the side that is used to calculate the area of the shape; any side can be considered the base **2.** for a prism, the two congruent, opposite, and parallel polygon faces that are used to name the prism; e.g., a rectangle-based prism has two bases that are rectangles. See *prism.* **3.** for a pyramid, the single polygon face that is used to name the pyramid; e.g., a square-based pyramid has one square base. See *pyramid.* **4.** for a cylinder and cone, the circle face(s). See *cylinder.*

benchmark measurement: a familiar measurement used to gain a sense of the size of a measurement or to compare measurements; e.g., 1 m is a common benchmark measurement, so a measurement such as 90 cm is more meaningful when you think of it as "almost 1 m"

benchmark number: a familiar number used to gain a sense of the size of a number or to compare numbers; e.g., 11 is just a bit more than 10 (10 is a benchmark number); also called an anchor number

capacity: a measure of the amount of liquid that a container shape could potentially hold; often measured in millilitres (mL) and litres (L); capacity units are also used to measure amounts of liquids; contrast with *volume*

cardinal number: a whole number that tells how many are in a group or set

category: when you sort objects or data into groups, the name of each group is a category; e.g., for a set of data about favourite pets, the categories might be dog, cat, and other

circumference: the name of the boundary of a circle; also the length of that boundary; also called perimeter

classify: to sort a set of items or data into groups and then name the groups to differentiate or categorize them; e.g., when you sort buttons into groups, you might classify them by colour and/or by number of holes

closed shape: a shape with a continuous perimeter, such as a polygon or a circle; contrast with open shape; e.g.,

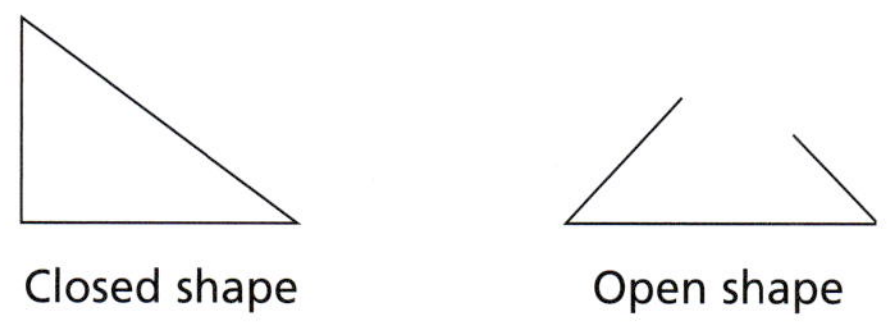

commutative property: a property of addition and multiplication whereby you can change the order of the numbers without changing the result, $a + b = b + a$ and $a \times b = b \times a$

composite shape: a shape that appears to be composed of multiple simple shapes such as triangles, squares, circles, and partial circles; e.g.,

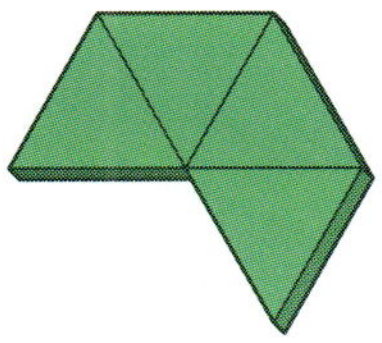

concrete graph: a data display made of the actual items or representative objects; e.g., a people graph of students in two rows to show whether they are wearing long or short sleeves (see **p. 123**), or a linking cube graph in two trains, one representing the students wearing long sleeves and the other representing the students wearing short sleeves (as shown below)

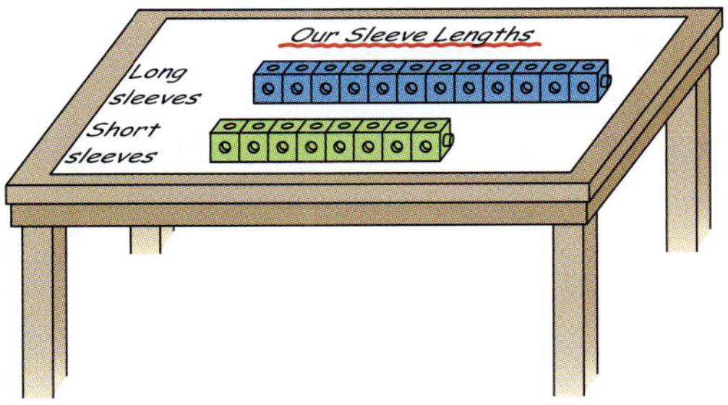

cone: a 3-D shape that consists of one circle base, one curved edge, one curved surface, and a point called the apex; e.g.,

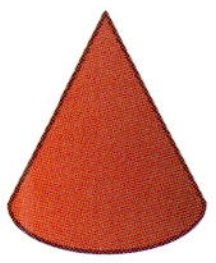

congruent: describes geometric figures that are equivalent; 2-D shapes or 3-D shapes are congruent if they have the same shape and size; e.g., these two triangles are congruent:

constant: a value that stays the same; e.g., the pattern 1, 3, 5, 7, … grows by a constant amount, 2

construct: formally, to draw using only straight edge (not a ruler) and compass and sometimes a Mira or geometry software; informally, to draw, create, or build a shape or graph

core of a pattern: the part of the pattern that repeats in a repeating pattern; e.g.,

Colour and shape core: blue-yellow and square-circle

corner: See *vertex*.

cube: a 3-D shape that has six congruent square faces; a cube is a special rectangle-based prism

curved edge: the place where the curved surface of a 3-D shape meets a face; e.g., a cone has one curved edge and a cylinder has two curved edges (see *cylinder*)

curved surface: a surface of a 3-D shape that is not flat; a sphere, a cone, and a cylinder each have one curved surface; e.g., see *cylinder*

cylinder: a 3-D shape that has two congruent, opposite, and parallel faces (bases), two curved edges, and one curved surface; e.g.,

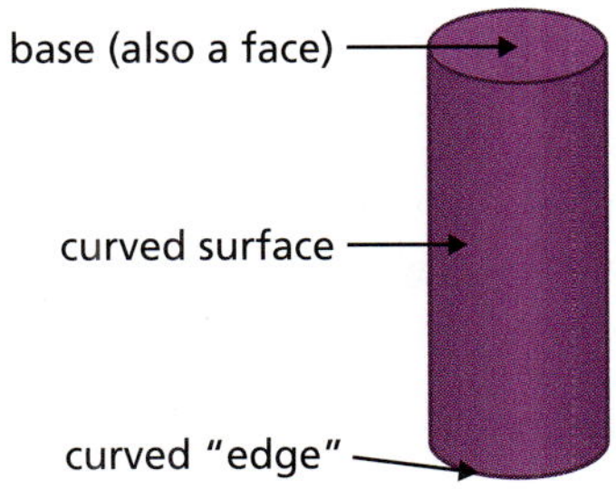

data value: a piece of data in a set of data; e.g., in the data set 3, 6, 9, 15, one of the data values is 9

denominator: the number below the bar in a fraction (it tells the number of equal parts in the whole); e.g., the denominator of $\frac{3}{4}$ is 4

difference: the result of a subtraction; e.g., in $12 - 7 = 5$, the difference is 5

digit: one of these ten symbols: 0, 1, 2, 3, 4, 5, 6, 7, 8, 9, used to create numerals; e.g., the digits 3, 4, and 5 in the numeral 345

dividend: the number you start with in a division (the number being divided into equal groups); e.g., in $24 \div 8 = 3$, 24 is the dividend

divisor: the number you divide by in a division (the number of equal groups or the number in each equal group); e.g., in $24 \div 8 = 3$, the divisor is 8

edge: a line segment on a 3-D shape where two faces meet; e.g., a triangle-based prism has nine edges (see *prism*)

elapsed time: a measure of the amount of time needed for an event to happen

element (in a pattern): a part of a pattern; e.g., in the shape pattern below, each shape is an element of the pattern; each number in a number pattern is an element of that pattern, although elements of number patterns are usually called terms

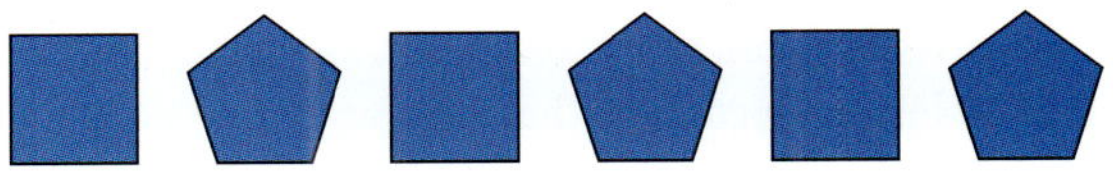

ellipse notation (…): a way to show that a pattern continues; e.g., 2, 4, 6, …

equation: a mathematical sentence that has an equals sign and may include one or more variables; e.g., $5 + 3 = 8$ is a numerical equation and $3 + \square = 5$ is an algebraic equation; an equation is an expression of balance, or the equivalence of two different representations of an amount

equilateral triangle: a triangle with three congruent sides; e.g.,

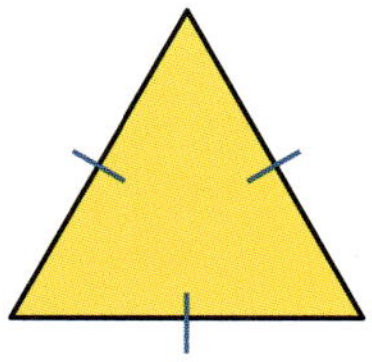

equivalent fractions: different fractions that represent the same part of the same whole; e.g., $\frac{1}{2}$ and $\frac{2}{4}$

Euclidean transformation: a transformation that does not change the size of a shape, such as a translation, reflection, or rotation

experimental probability: a way to describe how often something happens in an experiment or real life; it can be expressed using words or numerically; e.g., if the number 2 is rolled 3 times in 10 rolls, the experimental probability of rolling a 2 in that experiment is "not likely," or 3 out of 10. See *theoretical probability*.

face: a flat part of the surface of a 3-D shape; e.g., a cylinder has two faces and a triangle-based prism has five faces (see *cylinder* and *prism*)

fact family: a set of addition and subtraction equations or multiplication and division facts that share the same set of numbers; e.g., $2 + 3 = 5$, $3 + 2 = 5$, $5 - 2 = 3$, and $5 - 3 = 2$ is an addition/subtraction fact family

factor: **1.** in a multiplication sentence, one of the numbers being multiplied (the number of equal groups or the number in each equal group); e.g., in $3 \times 4 = 12$ (3 groups of 4 or 4 groups of 3), 3 and 4 are factors **2.** a number that divides into a whole number with no remainder; e.g., both 3 and 4 are factors of 12

figure: See *3-D shape*.

first-hand data: data that is collected by survey or observation; contrast with second-hand data, which is collected from another source that may or may not be first-hand data

flip: See *reflection*.

geometric attribute: a spatial attribute of a geometric figure, such as number of sides or vertices

geometric property: an attribute that all shapes in a certain classification share; e.g., all quadrilaterals have four sides, all rectangles have two pairs of congruent sides and four right angles, and all squares have four congruent sides and four right angles

geometric sequence: a number pattern in which the ratio between consecutive terms is the same; e.g., in the pattern 3, 6, 12, 24, 48, ..., the ratio is 2 : 1, or 2

hexagon: any six-sided polygon; a regular hexagon has six congruent sides and six congruent angles; e.g.,

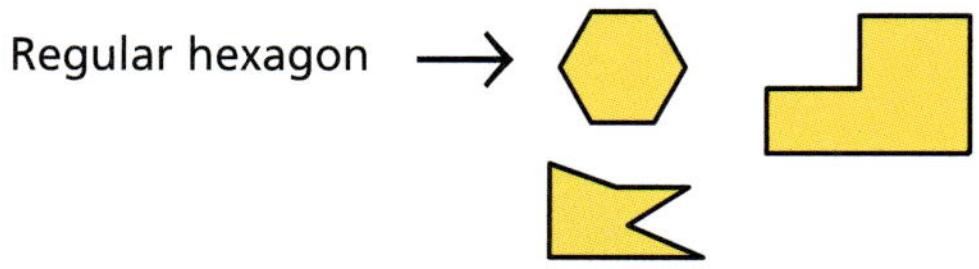

image: the result of a transformation. See *reflection, translation*, and *rotation*.

inequality: a mathematical sentence that has a less than or greater than sign and may include one or more variables; e.g., $3 + \square < 5$ is an inequality; an inequality is an expression of lack of balance or the non-equivalence of two different amounts

infer: to analyze a set of data and use reasoning to draw conclusions about the data

inverse operations: operations that undo each other; addition and subtraction are inverse operations, as are multiplication and division

isosceles trapezoid: a trapezoid (a quadrilateral with only one pair of parallel sides) that has two non-parallel congruent sides; e.g., a red pattern block is an isosceles trapezoid:

isosceles triangle: a triangle with two congruent sides; e.g.,

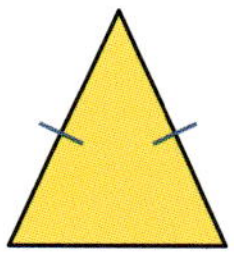

iterate: to repeat a uniform unit in order to measure something; e.g., you can iterate a craft stick along the edge of a table to measure the table's length in number of craft sticks, or you can iterate a square tile on the surface of the table to measure the table's area in square tiles

kite: a quadrilateral with two pairs of congruent adjacent sides and no parallel sides; e.g.,

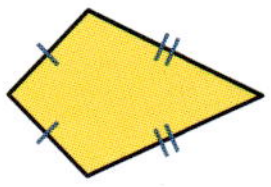

line of symmetry: an imaginary line that divides a 2-D shape in half such that, if you were to fold along the line, the halves would match; also called a mirror line; e.g.,

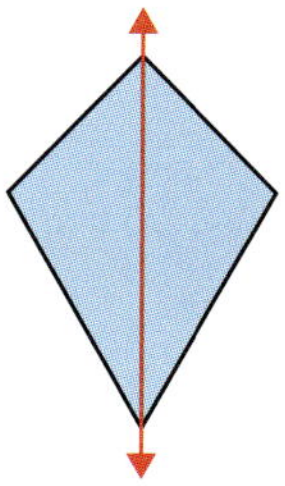

line plot: a way to organize, collect, and display data frequency in numerical categories along a number line base; e.g.,

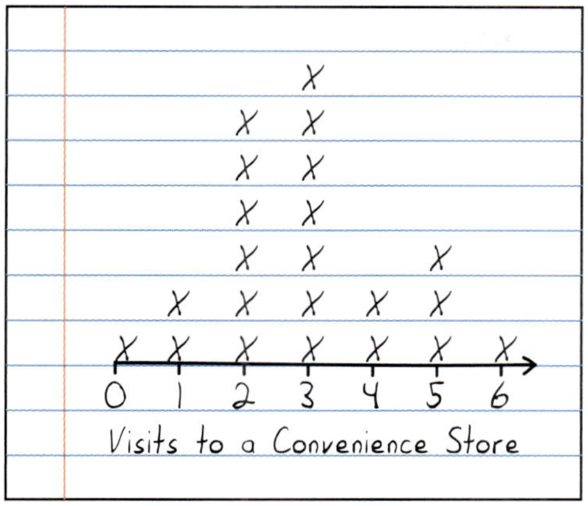

line symmetry: a 2-D shape has line symmetry if it can be folded in half along a line (line of symmetry) and the halves match exactly; also called reflective or mirror symmetry

mass: the amount of matter in an object, often measured in grams and kilograms; the mass of an object never changes, as opposed to weight, which is influenced by gravity

measurement attribute: an attribute of an object that can be measured, such as size, length, or mass

minuend: the number that you start with in a subtraction; e.g., in $15 - 12 = 3$, the minuend is 15

net: a model of the 2-D surface of a 3-D shape; if you assemble a net, it will form the 3-D shape; e.g.,

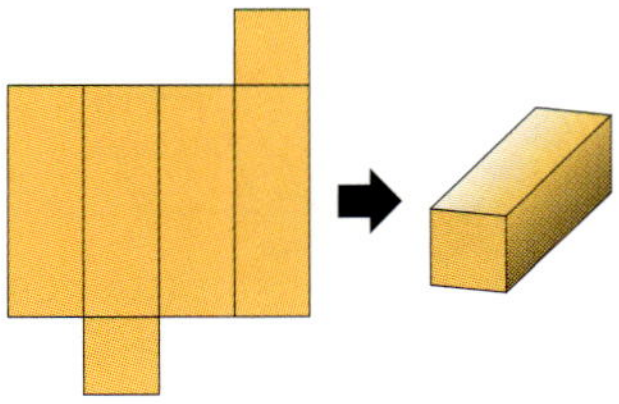

numeral: a symbol made of digits that represents a quantity (number); e.g., 15 is the numeral that represents the number 15

numerator: the number above the bar in a fraction; it tells the number of equal parts in the whole that the fraction describes; e.g., the numerator of $\frac{3}{4}$ is 3

object: See *3-D shape.*

open sentence: an early algebraic equation with a variable that is a blank or an open box; e.g., $\_\_ - 5 = 3$ or $5 + \square = 9$

open shape: See *closed shape.*

ordinal number: a number that describes position or order; e.g., in the pattern of shapes below, the third (3rd) shape from the left is a square

orientation: informally, describes a change in position of a shape; the mathematical definition has to do with the order of the vertices of a shape—when the orientation of a shape changes, its vertices are in a different order; a translation and a rotation do not change the orientation of a shape, whereas a reflection does; e.g., the reflection image below has a different orientation than the original shape. When you read the vertices in a clockwise direction starting at A or its image A′, the original shape reads ABCD but the image reads A′D′C′B′.

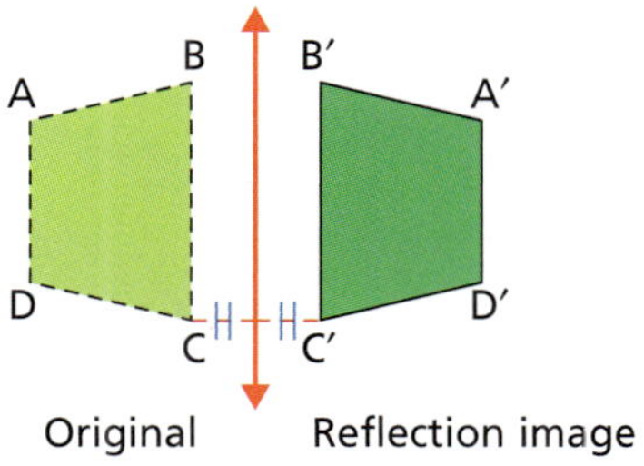

outcome: a result of an experiment; e.g., when you roll a die, there are six possible outcomes: 1, 2, 3, 4, 5, and 6

parallel: parallel lines and line segments will never meet when extended; e.g., the opposite sides of a rectangle are parallel line segments and the two lines below are parallel

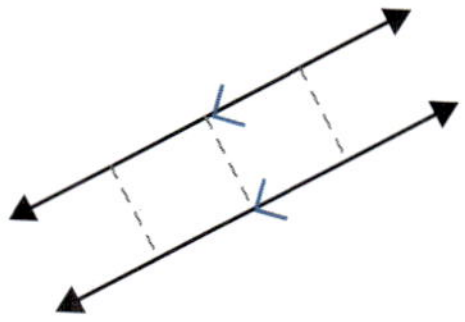

parallelogram: a quadrilateral that has two pairs of opposite congruent parallel sides; a square and a rectangle are special parallelograms

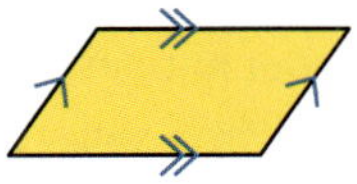

pattern rule: a description of a pattern that results in a unique pattern; it tells how the pattern starts and how it continues; e.g., the pattern rule for the growing pattern 2, 4, 6, 8, ... could be expressed as "start at 2 and add 2 each time" and the pattern rule for the repeating pattern 1, 2, 2, ... could be expressed as "1, 2, 2 repeating" or as "ABB, where A is 1 and B is 2."

pentagon: a five-sided polygon; a regular pentagon has five congruent sides and five congruent angles; e.g.,

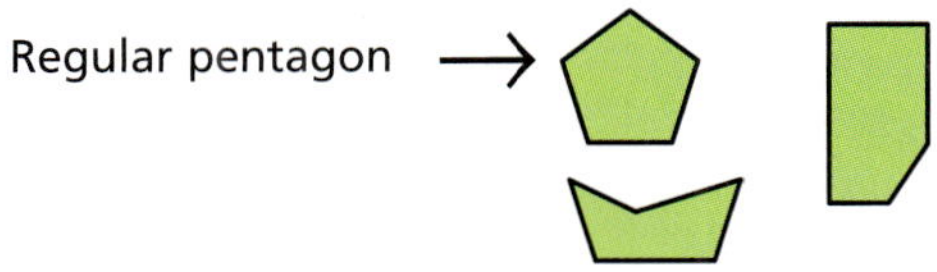

perimeter: the name of the boundary of a shape; it is also the length of that boundary (a linear measure); circumference is a special name for the perimeter of a circle

pictograph: a visual way to display data that uses the same picture or symbol throughout the graph; it represents a set of data that is grouped in categories; e.g.,

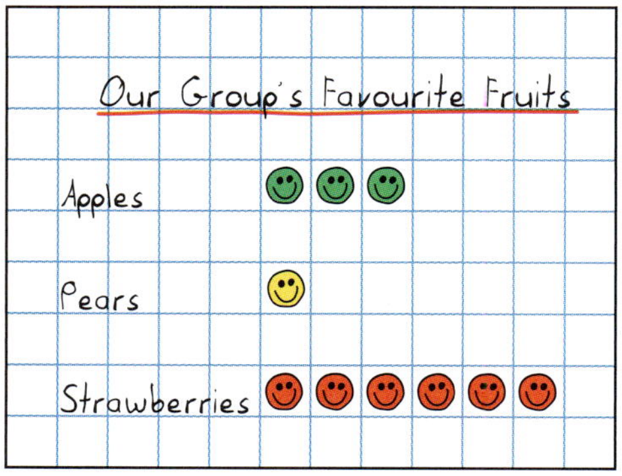

picture graph: a visual way to display data that uses a unique picture to represent each piece of data; e.g.,

place value: 1. a system in which the placement of a digit in a numeral determines its value; e.g., the 5 in 58 is worth a different amount than the 5 in 85 **2.** the names of the positions in a place value system; e.g., the tens place or the hundreds place

polygon: a closed 2-D shape with only straight sides; e.g., triangles, quadrilaterals, and pentagons are all types of polygons

polyhedron: a 3-D shape that has all polygon faces; e.g., prisms and pyramids are polyhedrons but cones and cylinders are not

prism: a 3-D shape with two opposite congruent parallel polygon bases and parallelograms (usually rectangles) for lateral faces; a prism is named by the shape of its base; e.g.,

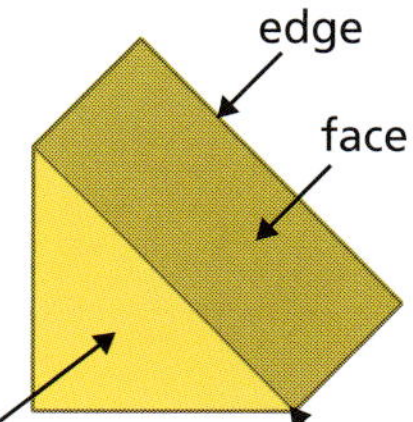

Triangle-based prism

product: the result of a multiplication (the total number in a set of equal groups); e.g., in $3 \times 4 = 12$ (3 groups of 4 or 4 groups of 3), the product is 12

property: See *geometric property.*

pyramid: a 3-D shape with one polygon base and three or more triangle faces; a pyramid is named by the shape of its base; e.g.,

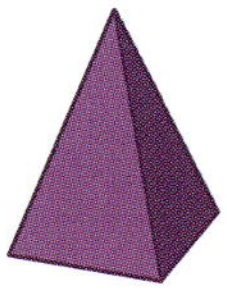

Square-based pyramid

quadrilateral: a four-sided polygon; e.g., squares, trapezoids, and rhombuses are all quadrilaterals

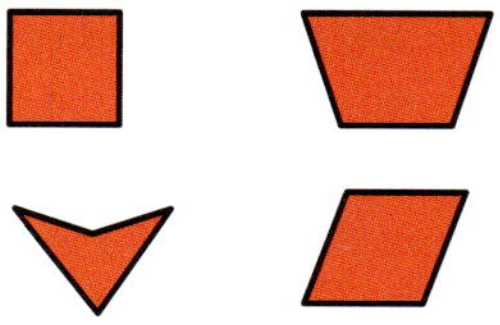

quotient: the result of a division (the number of equal groups or the number in each equal group); e.g., in $20 \div 5 = 4$ (20 divided in 5 equal groups or 20 divided into groups of 5), the quotient is 4

random: describes anything that happens by chance and is not predictable with certainty; e.g., the result of rolling a die or spinning a spinner

rectangle: a special parallelogram with four right angles; a square is a rectangle that has four congruent sides

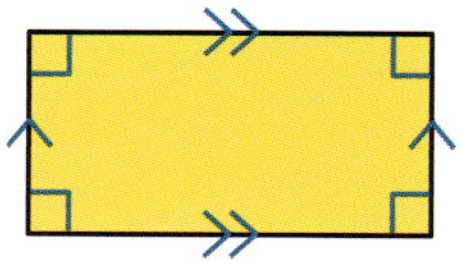

rectangle-based prism: a prism that has two opposite, congruent, and parallel rectangle bases; also called a rectangular prism; e.g.,

rectangular prism: See *rectangle-based prism.*

referent: a special benchmark measurement that is used to estimate other measurements; e.g., "if my little finger is about 1 cm across, then the book must be about 2 cm thick," or "I can visualize how long 1 m is, so I estimate the room to be about 5 m long"

reflection: a transformation that flips a shape across a reflection line without turning or sliding and results in a congruent mirror image with the opposite orientation; sometimes called a flip; e.g.,

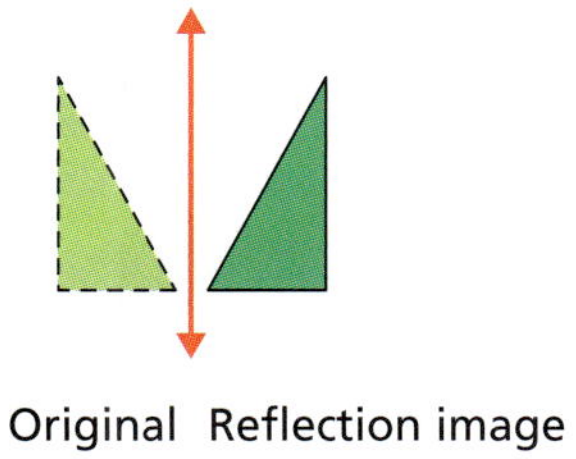

reflection line: the line across which a shape is reflected or flipped; sometimes called a flip line; See *reflection.*

regular polygon: a polygon that has congruent sides and congruent angles; e.g.,

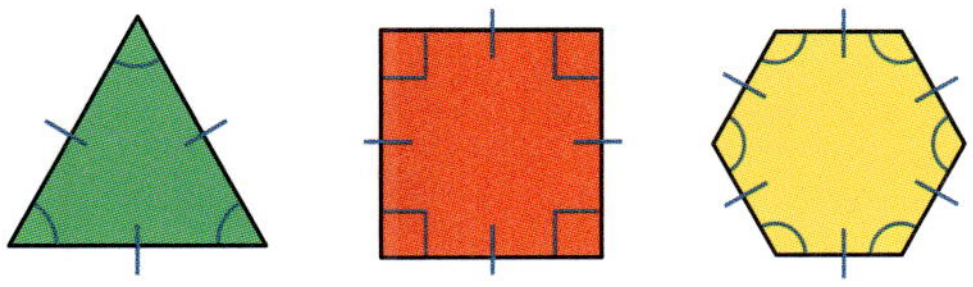

rename: **1.** to express a number in a different way; e.g., rename 12 as 10 + 2 **2.** to convert a measurement; e.g., rename 2 m as 200 cm

rhombus: a parallelogram with four congruent sides; a square is a rhombus that has four square corners

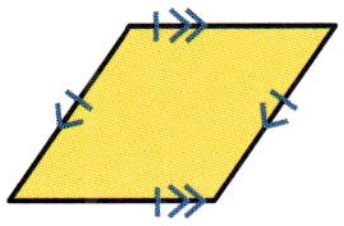

right angle: a 90° angle; informally called a square corner

right triangle: a triangle with one square or 90° angle; e.g.,

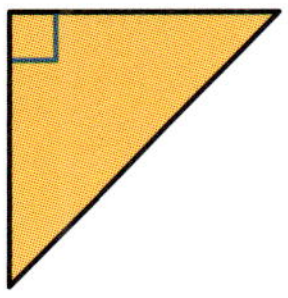

rotation: a transformation that turns a shape around a turn centre and results in an image that is congruent to the original shape with the same orientation; informally called a turn; e.g.,

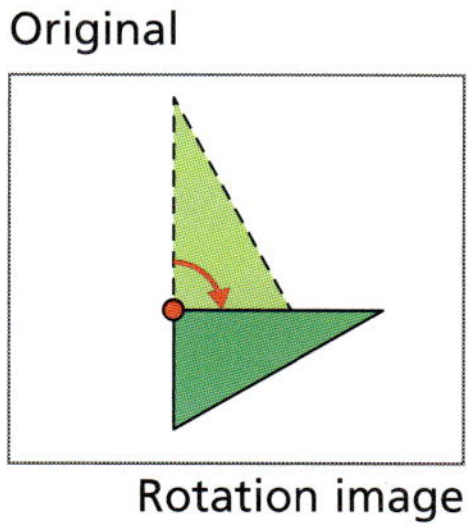

round: to estimate a number in a particular way to simplify its complexity for the purposes of reporting, comparing, or calculating; e.g., 24 and 13 rounded to the nearest ten are 20 and 10 if you follow the rounding conventions described on **page 34**, but to estimate the sum of 24 + 13, you might instead round to 25 + 15 because these numbers are comfortable to work with mentally.

scalene triangle: a triangle with three different side lengths; e.g.,

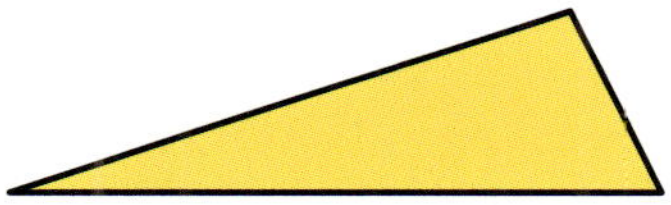

side: a line segment on a polygon; e.g., a triangle has three sides

skeleton: a model of a 2-D shape or 3-D shape that emphasizes certain components; in a 3-D skeleton, the vertices and edges are emphasized, and in a 2-D skeleton, the vertices and sides are emphasized; e.g.,

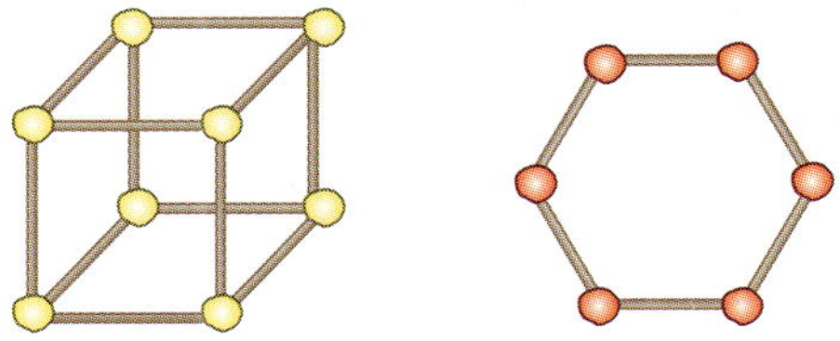

slide: See *translation.*

solid: See *3-D shape.*

solve an equation: to determine the value of an unknown in an equation; e.g., to determine the value of ❑ in 4 + ❑ = 10

sort: to put items into different groups according to a sorting rule. See *classify* and *sorting rule.*

sorting rule: a way to describe how a group of items have been sorted; e.g., the sorting rule used below is "shapes that are red"

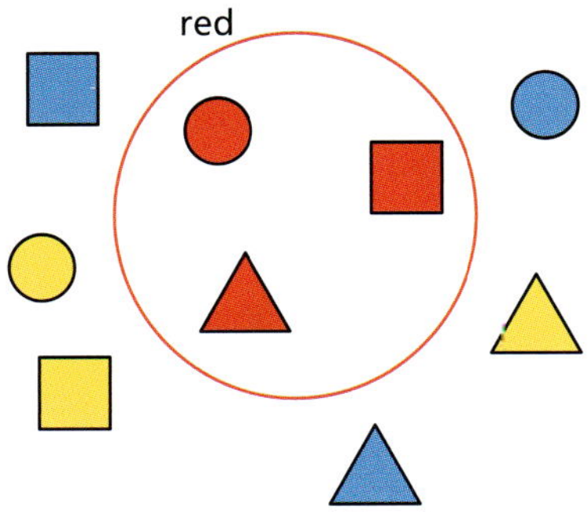

sphere: a 3-D shape in the form of a ball; mathematically it is defined as the set of points in space that are a constant distance from a fixed point, the centre

square: a rectangle with four congruent sides; e.g.,

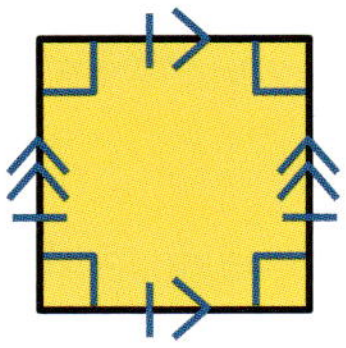

square-based prism: a prism with two square bases; the remaining faces are four congruent rectangles

square-based pyramid: a pyramid with a square base; the remaining faces are four congruent triangles. See *pyramid.*

square corner: See *right angle.*

standard form: the usual way of representing a number symbolically; e.g., 346 and 0.4 are in standard form

standard unit: a measurement unit that is defined, unambiguous, and universal; e.g., 3 m is a standard measure, but a measure such as 3 paces is not

subitize: to know how many items are in a set without counting

subtrahend: in a subtraction, the number that is subtracted; e.g., in $12 - 4 = 8$, the subtrahend is 4

sum: the result of an addition; e.g., in $2 + 3 = 5$, the sum is 5

tally chart: a data display used to describe the frequency of collected data in categories using a mark for each data value; these marks are grouped in fives for ease of counting; tally charts can be vertical or horizontal; e.g.,

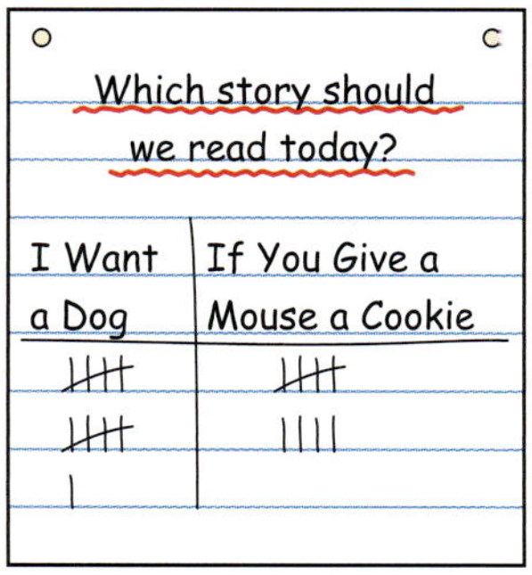

term: 1. each number in a number pattern is called a term **2.** in an expression, each part is a term; e.g., in 8 + ❑, there are two terms, 8 and ❑

theoretical probability: a number that compares the number of favourable outcomes to the total number of possible outcomes; e.g., the theoretical probability of rolling an even number on a die is $\frac{3}{6}$ or $\frac{1}{2}$ since there are 3 favourable outcomes (2, 4, 6) and 6 possible outcomes (1, 2, 3, 4, 5, 6). See *experimental probability.*

tile: to arrange 2-D shapes to cover an area without gaps or overlap; sometimes called a tessellation when all the shapes are congruent

transformation: a motion that moves each point in a shape in a consistent way, resulting in an image; a Euclidean transformation relocates a shape and may change its orientation (a reflection) but does not change it in any other way. See *Euclidean transformation, orientation, reflection, rotation,* and *translation.*

translate a pattern: to change a pattern to an equivalent but different pattern with the same underlying mathematical structure; e.g., the number pattern 1, 2, 1, 2, 1, 2, … is a translation of the shape pattern below because both follow an AB pattern

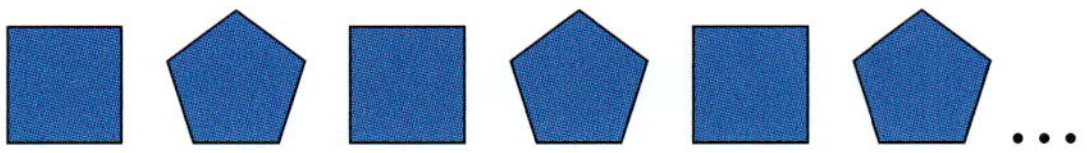

translation: a transformation that slides a shape without turning or flipping and results in an image that is congruent to the original shape with the same orientation; informally called a slide; e.g.,

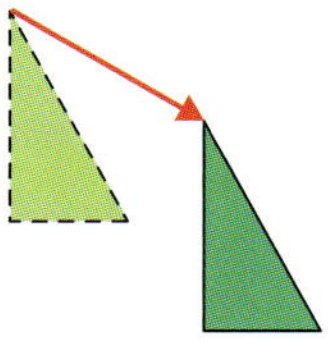

Original Translation image

trapezoid: a quadrilateral with only one pair of parallel sides. See *isosceles trapezoid.*

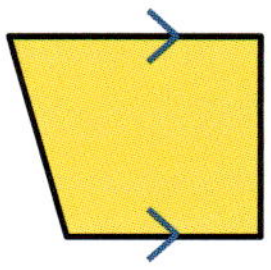

triangle-based prism: a prism that has two opposite congruent triangle bases, the remaining three faces are rectangles; also called triangular prism. See *prism.*

triangular prism: also called triangle-based prism. See *triangle-based prism* and *prism.*

unit fraction: a fraction with a numerator of 1; $\frac{1}{2}, \frac{1}{3}, \frac{1}{4}, \ldots$ are the unit fractions

variable: a letter or symbol that represents one or more unknown values in an equation or expression; e.g., in ❑ + 2 = 10, the variable ❑ represents 8; in the expression ❑ + 2, the variable represents the phrase "2 more than any given number, ❑"

Venn diagram: a visual system of one or more sorting circles that allows for cross-classification; e.g.,

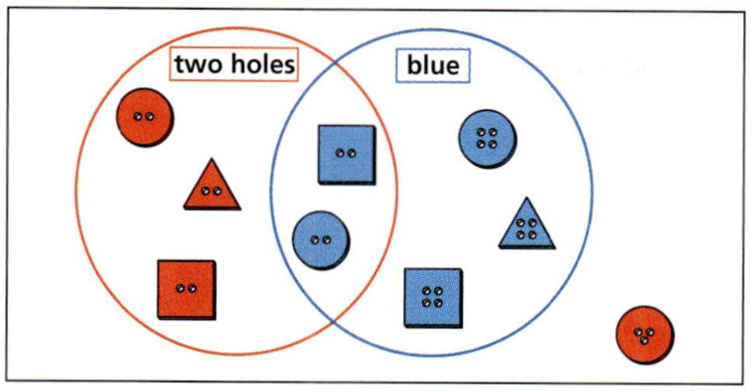

vertex: **1.** in a polygon, the point where two sides meet; e.g., a triangle has three vertices **2.** in a 3-D shape, the point where three or more edges meet; e.g., every rectangle-based prism has eight vertices (see *rectangle-based prism*); informally called a corner

volume: a measure of the amount of three-dimensional space that an object takes up; contrast with *capacity*

Index

N

O

P

R

S

T

U

V

W

Z